D1222741

Who Shall Take Care of Our Sick?

Medicine, Science, and Religion
in Historical Context

Ronald L. Numbers, *Consulting Editor*

# Who Shall Take Care of Our Sick?

## Roman Catholic Sisters and the Development of Catholic Hospitals in New York City

*Bernadette McCauley*

*The Johns Hopkins University Press*
BALTIMORE

Printed in the United States of America on acid-free paper
2 4 6 8 9 7 5 3 1

The Johns Hopkins University Press
2715 North Charles Street
Baltimore, Maryland 21218-4363
www.press.jhu.edu

Library of Congress Cataloging-in-Publication Data

McCauley, Bernadette.
Who shall take care of our sick? : Roman Catholic sisters and the development of Catholic hospitals in New York City / Bernadette McCauley.
p. cm. — (Medicine, science, and religion in historical context)
Includes bibliographical references and index.
ISBN 0-8018-8216-8 (hardcover : alk. paper)
1. Catholic hospitals—New York (State)—New York—History. I. Title. II. Series.
RA975.C37M38 2005
362.11'097471—dc22 2005000735

A catalog record for this book is available from the British Library.

# Contents

# Preface

As surprising as it may seem today, the hospital is a relative newcomer to American health care. Although pesthouses and military infirmaries date back to the seventeenth century, the first permanent hospital in what would become the United States did not open until 1752, in Philadelphia. Over the next 125 years, the number of institutions grew slowly. Unlike the hospitals of the twentieth century, these early American hospitals cared for few surgical or obstetrical patients. Rather, they provided simple care and shelter to the sick poor who had nowhere else to turn.[1] The first survey of U.S. hospitals, done in 1873, located only 178 nationwide. Then came the hospital explosion. During the next fifty years, the number of American hospitals increased dramatically—estimates reach as high as four thousand by 1900—and the hospital emerged as the central institution in medical treatment and training.[2]

In recent decades historians have begun to pay attention to American hospitals, and we now know a great deal about how the American hospital became the institution it is today. Two outstanding surveys of American hospital development and several excellent histories of hospitals in particular cities have been published, as have studies of specialized hospitals for groups such as African Americans, women, and mentally ill people.[3] Nevertheless, although religiously sponsored institutions made up nearly one-third of the hospitals in the United States in 1910, we still do not know much about these institutions. What is clear is that, of all the religious denominations operating hospitals in America, Catholics have been the most active. American Catholics opened their first hospital, Mullanphy Hospital in St. Louis, in 1832; by the end of the century they were running 10 percent of all the hospitals in the country.[4] As one nineteenth-century Catholic publication put it, it clearly had become "a matter of high policy, duty, and right, among Catholics to build and support hospitals."[5]

Catholics did so for numerous reasons, including fears that Protestant administrators and chaplains at other hospitals were not sympathetic to the religious needs of Catholic patients and that the Catholic clergy were not treated well or fairly in those hospitals. In New York, for example, Catholics complained that it was difficult for priests to visit the privately run New York Hospital. New York Hospital's rules were such that a patient had to request a visit from a clergyman; Catholic priests wanted to visit on their own initiative and find Catholic patients.[6] At Bellevue, the city's municipal hospital, Catholics protested that the Protestant chaplain was unfairly privileged with a salary. "We did not know Presbyterianism was the established religion at Bellevue," the Catholic *Freeman's Journal* fumed in an editorial in 1848, the year before the first Catholic hospital in New York City opened.[7]

Women play a prominent role in the history of Catholic hospitals. Like the overwhelmingly majority of Catholic hospitals in the United States, all of New York's Catholic hospitals were founded by women religious, properly called *sisters,* though commonly referred to (inaccurately) as *nuns.*[8] These women assumed primary responsibility for the nuts and bolts of Catholic hospital development. They raised money to initiate and maintain a hospital, managed it, set the standards of care, and provided the nursing. Sisters organized their hospitals to do more than create a nondiscriminatory environment for Catholic patients. Their purpose was also to advance their mission to the needy. They took the quality of their hospital care very seriously for two related reasons: because they believed that they had a spiritual mandate to provide good care and because they believed that their patients deserved it. For sisters, caring for the sick was a religious responsibility.

An apocryphal anecdote in the journal *Catholic World* at the end of the nineteenth century suggests, crudely but spectacularly, how Catholics promoted their hospitals. "A poor wretch was brought to a Sisters' Hospital and died after a few days of suffering. On entering he said he had no religion and no use for religion but the day he died he called for a chaplain. 'Sir,' he said, 'I want to die in the religion of that lady with the big bonnet taking care of me.'"[9] Catholic hospitals may have produced few conversions such as this, but they powerfully affected the lives of New Yorkers in other ways.

In 1849 the Sisters of Charity opened the first Catholic hospital in New York City. This was St. Vincent's. At the time the city already had two hospitals, but the sisters and their supporters believed they could do a better job than either the city administrators at Bellevue or the board of directors at the New York Hospital. By the time of the 1898 consolidation, which combined Brooklyn and New York, the Sisters of Charity had been joined in hospital work by Sisters of St. Joseph, Missionary Sisters of the Sacred Heart, Dominicans, and Franciscans—none of which had been organized primarily to run hospitals. Six years later a census revealed that Catholics were running fourteen of the city's privately run hospitals: seven general-care institutions plus specialized facilities for infants and children, for women, and for sufferers from contagious and incurable diseases. Among religious hospitals, Jewish hospitals came in a distant second with three; Lutherans were next with two. By 1924, the more than 4,500 beds in Catholic hospitals accounted for about one-quarter of all beds in the privately owned, nonproprietary hospitals in New York City. Sisters managed them all. At a time when women rarely occupied positions of leadership in business, scores of sisters sat on hospital boards and served as administrators of large and increasingly complex institutions. Twentieth-century sisters managed physical plants that included laboratories, kitchens, laundries, offices, emergency rooms, nursing schools, private rooms, and wards.[10]

In this book I seek to answer the following questions: What was distinctive about Catholic hospitals? Why did patients choose to go to them, and what kind of care did they receive when they did? How could patients tell they were in a Catholic hospital—by appearance, personnel, treatment? What role did Catholic hospitals play in the religious and medical worlds of New York City? How did Catholic involvement in health care influence Catholicism, and how did Catholicism influence health care?

The Catholicism of Catholic hospitals was certainly tangible. Sisters were ubiquitous, there were crucifixes in most every room, and most facilities had a chapel. Sisters never hid their religion, and they surely prayed for miracles, but their hospitals featured careful nursing and state-of-the-art medical care and treatment. How and why they did so is the focus of this book.

In examining Catholic hospitals in New York City from just before the Civil War through the 1920s, I explain why the sisters were so central to the development of Catholic hospitals and what made them unique among hospital nurses and administrators in New York. I look at the motivations driving the growth of Catholic hospitals, particularly the determination to establish a strong Catholic presence in an often hostile American environment. The specifics of Catholic hospital care in the latter nineteenth century are compared with that given at other hospitals and the sources of financial support for Catholic hospitals during that same critical period are analyzed. I then carry the story into the early twentieth century, when Catholic hospitals responded to the movement to standardize Catholic hospital management and care.

The most familiar Catholic institution in New York City is probably St. Patrick's, the city's Gothic-style cathedral on Fifth Avenue. Built to be more than a place to worship, the cathedral proclaims the Catholic Church's aspirations to power and prominence in nineteenth-century New York.[11] St. Vincent's Hospital, a contemporary of the Cathedral, while less imposing, shares its origins and a similar purpose. Both were the product of a contested social environment where Catholics sought to make their church a prominent force in the city's political and social landscape. Their founders also were related. St. Patrick's was the dream of John Hughes, the first Roman Catholic archbishop of New York, who was an outspoken leader of the American Catholic Church in the middle of the nineteenth century. Ellen Hughes, his sister, was St. Vincent's first administrator.[12]

Ellen Hughes never achieved her brother's notoriety, nor would she have wanted to. As Sister Angela, a member of the Sisters of Charity, she lived among other women similarly dedicated to a communal life of prayer and service to the poor. Her life and work were a quiet counterpoint to her brother's. John's style was aggressive; he addressed civic inequalities and the public manifestations of nativism head-on—from his pulpit, in the press, and at the ballot box. Sister Angela concerned herself with different kinds of issues facing immigrants, and she did so in less dramatic and confrontational ways. She is a lesser known historical figure, but the deliberateness of her efforts and her contribution to her church equal the archbishop's in their significance in New York City.

Sister Angela and a host of other sisters lightened the burden of illness and soothed fears about hospital care for several generations of immigrant New Yorkers, and they left the Catholic Church entrenched in the hospital landscape of New York City. They did so as women religious cohabiting the world of God and immigrant New York. They went back and forth between these worlds with ease. While sisters believed in miracles, they never promised them to their patients. Their treatment was not infused with zealotry but recognized that health care was more than medicine and surgery. The therapeutics of their health care was decidedly noncontroversial and up to date scientifically. Ironically, these otherworldly women established the Catholic Church's place in the very real world of mainstream medicine.

# Acknowledgments

It is a pleasure to thank everyone who has been so kind to me and supportive of this work. I begin my thanks at Hunter College with the many friends to whom my gratitude is deep and longstanding: Naomi Cohen, John Jones, Bob July, Mike Luther, Jo Ann McNamara, Naomi Miller, Bob Seltzer, and Nancy Siraisi, all of whom knew me both as a Hunter student and as a professor and have been supportive in a variety of ways. So have newer friends I met at Hunter especially Ann Cohen, Ira Cohen, Meg Crahan, Roger Persell, Polly Thistlethwaithe, and J. Michael Turner, without a doubt the best officemate anyone could have. Thank you also to the Hunter administration, and Acting Dean Judith Friedlander, for granting a leave that allowed me to complete final revisions and for the help of the wonderful librarians at Hunter who can find anything, especially to Danise Hoover, Clay Williams, and Patricia Woodard.

I begin a new paragraph by mentioning Barbara Welter because Barbara is in a category of her own. She is simply the best historian I know. She constantly reminds me why I came to study history in the first place. Her generosity is overwhelming, and it has been in this project, and in so many other ways, as Milo, Fiona, and dear Henry would also attest if they could. (Or have they?)

Barbara, Jane Cavolina, Venus Green, and Jack Salzman have each been very close to this book. They listened to me interminably, provoked me, consoled me, and kept the book and me on track. Where this book is good, they have been there. Where it is not, I probably didn't listen to them. That the manuscript made it to press is in no small way the result of Ron Numbers's support and tenacity, and my gratitude to Ron is equally tenacious. One of the nicest aspects of it all has been getting to know Ron and his family. Thanks also to Jackie Wehmueller at the Johns Hopkins University Press for sticking with us on this; I know her support was critical.

This work began under the direction of David Rothman, who pointed out where my real interest lay (that would be nuns), and with David Rosner, Elizabeth Blackmar, Patricia Byrne, and James Shenton, who helped me shape some thoughts and questions into the study that began this book. (David Rosner was less effective in describing himself as the tall guy with a beard and wearing jeans when I went to meet him at the coffee shop on Amsterdam Avenue near Columbia. If you have been there, you know what the problem was.) I am sorry Jim will not be here for a book party, where he could have held court.

The communities of women religious, whose history I examined in this book, were all extraordinarily welcoming, and the footnotes reveal that I could not have done this work without their collections. Many thanks to everyone in those archives and others who assisted me so graciously, in particular, archivists Sister Frances Maureen Carlin, O.P.; Sister Ann Courtney, S.C.; Sister Rita King, S.C.; Sister Margaret Quinn, C.S.J.; Sister Edna McKeever, C.S.J.; and Sister Mary Louise Sahm, S.F.P., who made my time with them such a pleasure. Thanks also to Msgr. Thomas Shelley for his quick and wise responses to last-minute New York Catholic questions and to David Thomas for some Welsh editing.

A fellowship year at the Center for the Study of American Religion helped me think about my research in different ways, and this book is better for it. Similarly, research funded by the Research Foundation of the City University of New York contributed to broadening my understanding of the topics addressed in this study.

All my friends, whether in the business or not, have heard way more than they probably needed to know about sisters and Catholic hospitals. Special thanks to Evelyn Ackerman, Marilyn Altman, Carla Bandini, Doron Ben Atar, Carol Candela, Jo Ann Candela, Charlotte Crystal, Babs Davy, Steve Deyle, Joe Diplasco, John Fider, Larry Fleischer, Rita Halbright, Jon Lory, Laurie Malinowski, David Mattern, Mary Frances McCarthy, Gemma Mulvihill, Anja Musiat, and Eamon O'Kane, who always asked how things were going and sometimes even took me in as I looked at yet another archive.

My family has also heard way too much but never complained. Thanks to Bridge and Boo and Mike and Phil for that and for lots of other stuff as well, especially Jamie, Michael, Jack, and Ben. Lucky for us all that my mother has always been enormously proud of us, as was my father. I know this book would have delighted him, and I treasure that.

Saving the best until last, I thank the most wonderful person I know.

Anyone who knows me knows just who that is and, moreover, knows that thanking him for smiling at me while I finished this book is just like tapping at the tip of the iceberg that brought down the Titanic. (The reference is not inappropriate; Titanic survivors were brought to St. Vincent's.) But I close here with many smiles of my own, all thanks to George.

CHAPTER ONE

# "A Climate New to Them"
## *The Foundations*

When the first Catholic hospital opened in New York in November 1849, it was only the city's third hospital and was the first to be organized by a religious group. Like the other two (Bellevue and the New York Hospital), St. Vincent's was a general care hospital; it accepted patients of both sexes, of all ages, and those suffering from a variety of diseases and conditions. It was the first of numerous Catholic hospitals that would flourish in the following decades, mainly between 1870 and 1900, in Manhattan and surrounding areas.

How and why this came to be is a story about choices made by the Catholics who organized hospitals and their supporters who, as fundraisers and patients, kept the hospitals in business. While the founders of Catholic hospitals cared about therapeutics, Catholic hospitals were not organized to promote specific medications or clinical treatments or to introduce any Catholic religious ritual associated with healing. Rather, the founders wanted to institutionalize medical treatment that infused standard medical practice with a Roman Catholic perspective on life and death. Sisters' efforts and their interest in health care, at St. Vincent's and other hospitals, were a manifestation of the basic tenets of their lives as religious women in the context of immigrant life in New York City in the nineteenth century.

St. Vincent's was not the first Catholic hospital in the United States. Mullanphy Hospital in St. Louis (organized in 1829 and completed in 1832) had that distinction; another in Buffalo, which opened in 1839, was the first in New York State. As in the rest of the country, most other Catholic hospitals in New York were founded after the Civil War.[1]

The church's first involvement, or more accurately the earliest Roman Catholic interest, in health care in New York City predated hospital development and focused on the city's public hospitals, those managed by city authorities. They were the very first hospitals in both

*Fig. 1.* There is little in this portrait of Mother Jerome Ely, S.C., to suggest her work, but she and other sisters were a familiar sight among immigrant Catholics. Ely was the administrator at St. Vincent Hospital in Manhattan from 1855 to 1861 and, at different times in the century, mother general of the Sisters of Charity of New York. Photograph ca. 1875; Sisters of Charity of New York

New York and Brooklyn and began as extensions of city almshouses. These were places of last resort that housed destitute people, including many sick people who ended up there because they had no money to treat their illnesses. Bellevue, Manhattan's first city hospital, was officially separated into an almshouse and hospital in 1849. The number of public hospitals increased with the city's population. Except for Bellevue, most of New York City's public welfare institutions in the nineteenth century were located on the islands in the East River across from Manhattan. At different times, these islands housed hospitals for children and for those suffering from chronic and contagious diseases. In Brooklyn, similar city-run facilities existed in Flatbush.[2]

Eligibility for treatment at a city institution was based on the level of care deemed necessary by city officials, a residency requirement, and financial need, although it is not clear how rigidly procedures were followed. Regulations in Manhattan in the 1880s specified that an individual had to have lived in the city for one year and been approved by a local charity officer before he or she could receive treatment. According to the stated requirements, "Invalid applicants . . . must be provided with a permit, good for five days . . . giving name, nativity, age, occupation and residence in the city. It must be shown that the applicant is entirely destitute. The permit is delivered to the warden of Bellevue, the diagnosis of the disease is made by the examining physician and the patient assigned to the correct hospital." Recent immigrants who did not meet the residency requirement were eligible for treatment at the Emigrant Hospital run by the Emigration Society and could apply at Castle Garden for admission to that hospital on Ward's Island, located to the west of Manhattan in the East River. (Castle Garden was the point of debarkation for immigrants until Ellis Island opened in 1892.)[3]

By midcentury, the patient population at city hospitals was overwhelmingly foreign-born. Between 1849 and 1859 more than three quarters of the patients at Bellevue were immigrants. By 1866, more than half the admissions had been born in Ireland. The immigrant nature of the institution would probably have been even more obvious if the native-born patients had specified the nativity of their parents.[4]

Not surprisingly, the hospitalization of Irish immigrants at the public's expense attracted attention—native-born New Yorkers worried about how much all this charity was costing them. Recognizing this, many health reformers used the cost of charity to support their proposals for preventive measures. While some publicized health statistics to suggest the need for improved city sanitation, others simply looked at the figures and blamed the victims. The amount of public money that went to support the institutionalized immigrant became a popular target for attacks on immigrants.[5]

Although the Irish were not the only immigrants in New York, many considered them to be the most different and troublesome because of their religion and poverty. Fears and concerns about the Irish, many of whom were clearly in dire straits, were directed toward the Catholic Church, which some Protestants accused of not taking care of its own. The Association for Improving the Condition of the Poor, for example, contrasted the Roman Catholic Church with the city's Protestant churches. Its 1856 annual report explained that "all of our Protestant churches are charitable institutions," but the Catholics "make no corresponding provision for their poor."[6]

While New York Protestants worried about Catholics and the health of the city, the Catholic Church was complaining about the treatment of Catholic patients in the city's public hospitals. In much the same way they objected to Protestant involvement in public schools, the Catholic hierarchy feared the power of the Protestant churches in hospitals and worried about the influence of the Protestant clergy on the needy immigrant hospital population.

Protestant interest and involvement in the city's charity institutions began in 1785 when the municipal government authorized Protestant clergymen to preach in the Bellevue almshouse. In 1812, the Interdenominational (Protestant) Society for Supporting the Gospel among the Poor of New York was organized to conduct services there. This group received the financial support of the city through salary grants made to their designated minister.[7]

The Society for Supporting the Gospel Among the Poor was anxious to place its chaplain at Bellevue because, like other nineteenth-century reformers, its members saw the lack of religion as a primary cause of illness and dependency. As Charles Rosenberg has shown, reformers and religious leaders emphasized the connection between the inmates' spiritual and physical degeneration. In the words of one hospital chaplain, "All the bad diseases, or Nine out of 10, are produced by bad habits—or rum."[8]

The role of the Protestant chaplain in a city hospital was not simply one of religious convenience but was an integral part of treatment. The Bellevue chaplain played an active part in the operation of the institution. In the annual report of the Almshouse Commission in 1848, for example, his comments are included along with those of the superintendent and resident physician. He describes his weekly visits to the hospitals where he "leaves no room unvisited" with help from "City Missionaries" and "two Christian men" who distributed tracts, read the scriptures, and made themselves available for "religious conversation and prayers with and for the people in their state of affliction."[9]

Like their Protestant colleagues, New York's Catholic clergy also recognized a connection between illness and the religious life of the immigrant. As early as 1834, Bishop John Dubois attempted to open a Catholic hospital in New York because of what he saw as the overwhelming physical and spiritual needs of new immigrants, specifically the poor Irish ones. Trying, unsuccessfully, to solicit funds in Europe for this hospital, Dubois explained it would offer them "the necessary relief, attendance in sickness, and spiritual comfort, amidst the diseases of a climate new to them." The conditions the bishop referred to involved more than New York's icy Hudson winds. Dubois and his successors were very concerned about the religious climate of New York City and feared the activities of the Protestant churches within the city's public charities.[10]

The Catholic hierarchy was anxious to place Catholic clergymen in the early public institutions. Priests from the nearby parish of St. Stephen's visited Bellevue as early as 1828. The Jesuit Fathers at St. Francis College ferried across to the island institutions beginning in the late 1840s, and some eventually took up residence there. Their visits were not without problems. There were disagreements between the priests and hospital administrators over what a priestly visit could include and how long it could go on. The Jesuits complained that since city officials considered the Roman Catholic sacraments to be idolatrous rites, they were allowed to visit the hospitals but not to administer the sacraments.[11]

For most of the century, Protestant and Catholic clergy competed for the religious life of the patients at the city hospitals. Comments of the Protestant chaplain at Bellevue in 1848 reflect the animosity between them. Rev. Lyall complained that Roman Catholic priests "give their influence against the reading of our Bible—supplying none of their own, that I have ever seen; and one of them has shown decided hostility especially to tracts."[12] At the same time, the Catholic *Freeman's Journal* voiced a complaint about Protestant chaplains in some hospitals. "He should not be permitted to force his opinions down the throats of Catholic patients, as if it formed part of their medical treatment, and also Catholic inmates and other inmates who wish to have the assistance of a particular clergyman should not only be permitted but aided to do so."[13] While all New York's clergy seem to have agreed on the need for religious influence within the hospitals, they obviously disagreed on who should be supplying it.

The question of salary was another area of contention between Catholics and Protestants. Beginning in 1848, Catholics made repeated attempts to have Catholic chaplains paid salaries at city hospitals, petitioning the board of alderman "to have the Catholic clergymen attending Bellevue Hospital, paid."[14] The issue was still being discussed ten years later when the pastor at St. Stephen's wrote the archbishop that he was not "very sanguine in this regard."[15]

In one case, the administration at a city hospital lowered the salary it was paying its chaplain, a Protestant minister, noting that the Catholic priest who visited did so at no cost to the hospital. That chaplain left and an Episcopalian minister, whose church paid his hospital salary,

*Fig. 2.* Sisters of Saint Joseph at Saint John's Hospital Long Island City: Sisters Mary Laurentia Bradburn, Saint Philip Collery, Mary Joanna Ferricks, Marion Gannon, Saint Ludwina Johnston, Mary Catherine Molloy, Mary David O'Brien, Mary Denis O'Connor, Mary Hilda O'Malley, and Francis Clare Snyder. Those in the back row wearing caps are lay sisters; seated front and center is the hospital's founder and superior, Mary David O'Brien, to whom the bishop wrote of his "surprise" that the hospital debt was paid off. O'Brien was Irish-born and entered the Sisters of Saint Joseph convent in Flushing in 1873. She died at St. John's in 1904. Photograph ca. 1902; Sisters of Saint Joseph, Brentwood, New York

took his place. Perhaps in an attempt to avoid a similar situation, other Catholic requests for salary mentioned no desire to interfere with the current chaplain's salary, which, as one Catholic priest noted, "was meager enough."[16]

The Catholic hierarchy focused its attention on the salary question for more than financial reasons. Catholics felt the discrepancy reflected a fundamental difference of opinion over the status of the religion of patients in city institutions. In the eyes of the Catholic Church most of those immigrants were Catholics because they had come to the United States from a traditionally Catholic culture. Because of the overwhelming number of Irish immigrants and, to a lesser extent Germans, in city charity institutions, the Catholic hierarchy agitated for an active Catholic presence. Catholics felt that if any chaplain were to receive a city salary, it ought to be the Catholic clergyman since the work to be done was really his. Comments in the Catholic press about the paid Presbyterian minister at Bellevue made the Catholic position clear. Noting that "we did not know before that Presbyterianism was the established religion at Bellevue," Catholics complained that while it "is amongst the truest acts of charity toward these poor sufferers that they should have the services of their respective clergymen . . . if there is any salary to be given it certainly should not be given to the one who has the least work."[17]

Of course, New York's Protestant churches saw the immigrants and their religious status in a completely different light. They were, in the eyes of one visitor to Bellevue, "Irish of the most common sort." Patients were only nominally Catholic because "many of them could barely be called Christians."[18] Most important, they were "very accessible to kind words, and many of them will read what we put into their hands."[19] In other words, they were potential Protestant converts.

The Catholic hierarchy also complained about its clergy's treatment at the privately operated New York Hospital, the only other hospital in New York City before St. Vincent's opened. New York Hospital also had a Protestant chaplain who was paid a salary. Roman Catholic clergy were able to visit the hospital in the antebellum period but not easily. In 1851 the hospital established new rules for visiting, requiring a patient to request to see any other than the official clergyman before a visit could take place. The archdiocese complained about the new rule. As Rev. James Roosevelt Bayley, secretary to the archbishop, wrote to the board of governors, "the Catholic clergy of the City are very few in proportion to the work they are obliged to do, and if the Priest who attends the hospital was obliged to go to it, every time that one of the patients needed his services, he would have to visit it several times the same day." Catholic attempts to have the rule changed were unsuccessful as the board deemed it "inexpedient to make any change in the existing regulations for the house on the subject which allows a patient to send for any Minister that he may prefer."[20]

Roman Catholics were not alone in their complaints to New York Hospital. Other requests for a more open visiting policy came from a tract organization and the Methodist Episcopal Church, but there was a difference between these requests and the Catholic petition. As in their protests regarding city institutions they noted that the Catholic clergy should have a special position since "a majority of the usual inmates are Catholic."[21]

Both Catholic and Protestant chaplains at Bellevue and the New York Hospital were correct in their assessment of the religious status of the patient population. The immigrants who were filling up hospitals over most of the nineteenth century were Catholics by birth but not by practice. Jay Dolan's research on New York City's Irish and German parishes in the years between 1815 and 1865 indicates that a great many of New York's Irish, and some Germans too, were Catholics in name only. Dolan found that many immigrants chose not to attend Sunday Mass or even marry in the Catholic Church.[22]

The Catholic clergy was aware of the number of less than rigorous Catholic immigrants, although some might have been reluctant to admit it. New York monsignor and diarist Richard Burtsell recorded a conversation he had in 1865 with another priest on immigrant religious habits, where it took some doing for Burtsell to convince his colleague "that half of our Irish population is Catholic merely because Catholicity was the religion of the land of their birth."[23] Burtsell saw little improvement in the situation over the years as he and other colleagues tried to persuade New York's second archbishop, John McCloskey, that the church needed to do more to reach the large number of Irish Catholics who had little or no contact with the church.[24]

Church officials often blamed the problem on a shortage of priests and churches. Burtsell mentioned that if New York had more priests, they could "rake up those who by neglect have grown careless."[25] More was involved in this issue than mere numbers, however. The crucial factor was the nature of the religion practice the immigrants brought with them to the United States. Many early-nineteenth-century Irish immigrants came from an environment where religious observance and responsibility was often slight or nonexistent. As Jay Dolan explains, "It is clear that all Catholics did not come to the United States in sound spiritual condition. Many had not regularly attended worship services in Ireland, and others had not received the sacraments of confession or communion for years. In their adopted homeland such habits were not quickly discarded."[26]

Dolan estimates that only half of New York's Irish population at midcentury was an active part of the Catholic Church. The other half "lived on the fringe of parish life." They "were the anonymous Catholics" who left behind very little record of their religious lives because the parish church was not a fundamental institution for them. As Dolan concludes, "It was only one institution in the neighborhood, and in the antebellum period it attracted a limited percentage of newcomers."[27]

The church's concern for hospital visiting privileges reflected fears that the Protestant chaplain at Bellevue was correct in suggesting that these immigrants could become Protestants. Catholic clergy worried, with good reason, that a hospitalized Irish immigrant might not ever call for a priest, even if it were allowed upon request. Rev. Burtsell recorded at least one such unsuccessful sick call noting, "A dying Catholic acted rather obstreperously: was not very anxious to receive the sacraments."[28]

Visiting the sick was just the first step in an attempt by the Catholic Church to gain a position equal to that of the Protestant churches in charity hospitals, particularly the ones managed by public authorities. Fears about the fragility of the immigrants' faith encouraged the church to continue to maintain an active and visible presence on hospital wards.

Some clergy were optimistic about the potential for bringing fallen Catholics, in a hospital, back to the fold. One Jesuit referred to the city's charity institutions as "a royal hunting ground."[29] Another noted, "Persons are constantly met in the Hospital . . . who have never made their first Communion, not even their first Confession, or who have almost entirely forgotten what religious knowledge they may have acquired in their youth."[30] More realistic priests recognized that this could be a formidable task: "Many want to die as Catholics," observed one priest, "but they don't want to live that way."[31]

The position of the Catholic Church within the city's public charities improved markedly during the Civil War. According to one priest who visited city-run institutions, by 1861 the prejudice against Catholic priests had "yielded or was forced to yield" because of the tenacity of the clergy.[32] That same year Bishop Hughes commended the Commissioners of Charity and Health, the municipal board responsible for the city's public hospitals, for their "true impartiality and fairness which places all religion on a perfect equality."[33] By 1863, the *Metropolitan Record,* for a time the unofficial voice of the archbishop, acknowledged in an editorial that religious liberty was now a reality at the city charity institutions and "the fact that anything like religious distinctions are completely ignored, shows that a complete and beneficial change has been effected."[34] While the presence of Catholic clergy might have still disturbed some hospital authorities, as another priest on Ward's Island

*Fig. 3.* The pharmacy at St. Vincent's Hospital in Manhattan in 1904 as pictured in the annual report for that year. Sisters were most usually touted by their supporters for bedside nursing, but they worked in other capacities in their hospitals too. Sisters of Charity of New York

noted in 1872, "once established there none of the Commissioners had the courage to send me away."[35]

In the last quarter of the century, some Catholic chaplains began to receive salaries from public authorities. At the Brooklyn City Hospital both the Catholic and Protestant chaplains received a salary of $300 in 1887. Some of the Jesuits at the island institutions were paid salaries in 1890, and there is some evidence that the Catholic chaplain at Bellevue received a salary in 1889. The New York State Freedom of Worship Act of 1892, which acknowledged the free exercise of religion within any government institution, was the final legal step to providing Catholic clergy equal status under the law.[36]

While Catholic clergy were making significant steps in city-run institutions, the Catholic Church in New York continued to open its own hospitals. The majority of the city's Catholic hospitals were founded in the late nineteenth century, even as those religious restrictions at public institutions were lessening and priests had more access to patients. The second Catholic hospital to open in New York was St. Francis' Hospital, which was organized in 1865. The hospital was founded as much in response to circumstances at St. Vincent's, which was so much part of the Irish immigrant world of New York City, as to those at the city hospital, Bellevue. St. Francis' was intended specifically for German Catholic immigrants. St. Vincent's had not been founded as an "Irish" hospital, but the Sisters of Charity who ran the hospital were overwhelmingly Irish—and so were the patients they cared for.

Other Catholic hospitals would have similar ethnic connections. The Sisters of Charity remained closely connected to Irish New Yorkers and so did their hospitals. St. Catherine's in Brooklyn was founded in 1871 through the efforts of a German parish there by Dominican Sisters originally from Germany; Columbus Hospital in Manhattan opened in 1892 under the direction of an Italian immigrant sister, Frances Cabrini, and was organized for the care of Italian immigrants.[37]

These ethnic hospitals were an extension of successful efforts by national groups to organize separate parishes within the dioceses of New York and Brooklyn, and they reflected the cultural differences among Catholics. Religious traditions and practices varied among nationalities, so German (and later, Italian) Catholics in New York were anxious to organize their own churches. Irish priests dominated Catholic New York in the nineteenth century, and these other Catholic groups looked for some autonomy and the opportunity to create a religious environment in the style of their homeland.[38] The first German parish in the Archdiocese of New York was St. Nicholas Church, which opened on East Second Street in 1833. Brooklyn's first German parish was Most Holy Trinity Church, founded in Williamsburg in 1841. New York's first Italian parish, St. Anthony of Padua, was established in Greenwich Village in 1866, but most of the city's Italian national parishes were organized later in the century as the city's Italian immigrant population began to increase substantially.[39]

Language was important in the movement to organize both ethnic parishes and ethnic hospitals. When German Catholics first requested a church of their own in New York City they cited the need for a priest "who is capable of undertaking the Spiritual care of our souls in the German language."[40] Italian Catholics also complained that parish priests who did not speak Italian could not adequately serve Italian parishioners. Notably, Italians at one church in Manhattan complained that the priest assigned to their care could not make sick calls to them because he did not speak their language.[41]

The founders of several Catholic hospitals were anxious to provide physical care in the context of the patient's native language. The administration at Columbus Hospital, for example, noted that even in the best of circumstances, "our poor Italians . . . were not able to make themselves understood."[42] The immigrants' unfamiliarity with English was viewed as a problem with serious implications. The priests instrumental in the organization of St. Francis' feared that because of the language barrier, "the sick and infirm of the Congregation . . . were not satisfactorily well cared for in public hospitals."[43]

Catholics opened their own hospitals as a response to other kinds of abuses at municipal institutions too. In the decade following the Civil War, large municipal hospitals, Bellevue in particular, were often criticized by reformers for filthy and unhealthy conditions. Visitors in the 1870s reported on the intolerable state of affairs there, concluding that "Bellevue was a very much mismanaged institution; three patients sometimes slept on two beds, five patients on three beds, and it happened now and then that they slept on the floor. During two weeks in January, 1876, there was no soap in the hospital, and not enough clothing; many patients had neither pillows or blankets, and forty-eight percent of the amputations made proved fatal."[44] Reformers further noted the irony that "the most frequented refuge of the sick in this great city is notoriously liable to the suspicion that it does harm to those who are brought within its walls."[45] The Brooklyn public hospital at Flatbush was similarly described, with complaints of "poor food, scant clothing, indifferent nursing."[46]

Catholic hospital founders offered the possibility of better medical care than was available at the city-operated hospitals. Yet even amid criticisms like those noted above, Catholics overwhelmingly cited religious reasons when they explained the need for Catholic hospitals. The hierarchy did voice some concern in 1850 about the lack of Catholic physicians at the city hospitals, but that point was never pursued, probably because there were few Roman Catholic doctors in New York at that time. Jay Dolan's sample of the occupations of the parishioners at midcentury immigrant parishes reveals few physicians among them.[47] (The first president of St. Vincent's medical board and the chief surgeon, for example, was New York's leading surgeon, Valentine Mott. Mott was also chief surgeon at New York Hospital and Bellevue and was not a Roman Catholic.)[48]

Although Catholics did not dwell on the need for Catholic physicians, they did talk about the comforts of Catholicism in a hospital. A fictional account of a Catholic hospital published in the Catholic press in 1862 describes the experiences of a hospital nurse named Sister Magdalen and a Protestant patient. From the beginning of the story it is clear to Sister Magdalen, and to readers, that the patient is not going to survive. The hospital's medical capacity is never at issue; what is important is Sister Magdalen's attention to the dying man and his family.[49]

The nurse is unable to help the inconsolable wife who keeps a lonely vigil beside her dying husband's hospital bed. Because the woman is a Protestant, Magdalen cannot comfort her with the suggestion she would offer a Catholic in a similar situation: to compare her grief to that of Mary watching Christ suffer. "There was a cloud which obscured from her the cross of Jesus and the heart of Mary, the Catholics' great consolation and refuge." But Magdalen knows that she can still "pray for them."[50] Throughout the patient's final hours she kneels by his bedside with his wife and mother where she "forgot I was praying by a Protestant deathbed; and . . . invoked the aid of Mary all powerful." Before the patient dies he temporarily regains consciousness and, with Sister Magdalen's prompting, peacefully leaves this world with the words, "Jesus receive my soul" on his lips. The story continues after his death: the grief stricken wife falls ill with a fever and Magdalen nurses her for several weeks. When the grateful widow recovers, she announces that she wants to become a Sister of Charity, just like her devoted nurse. She converts to Catholicism and the piece ends happily as she enters the convent.[51]

This parable shows how proponents depicted the special and, in their eyes, superior nature of Catholic hospital care. The story also demonstrates that medical treatment was not their only priority. Although the patient dies, this operation is clearly a success. Although highly romanticized, Sister Magdalen's story represented a very real expectation that Catholic hospitals would be a more comfortable place to be in sickness and in death, and could increase the visible ranks of the faithful.

The rhetoric surrounding initial efforts to open Catholic hospitals mirrored the church's earlier arguments in support of Catholic schools. During a highly publicized controversy over public funding of schools in New York in the 1840s, Bishop Hughes made it clear that although he objected to the religious orientation in the city's public schools because it was Protestant, and often antagonistic toward Catholicism, he was not lobbying for the removal of religion from the curriculum. Indeed, he found the idea of nonsectarian education just as alarming than a sectarian kind that discriminated against his own.[52] Similarly, advocates of Catholic hospitals in the nineteenth century commented on the lack of religious influence in public hospitals. They argued that even in the best possible circumstances—no Protestant proselytizing—those hospitals offered inadequate care to Catholic patients. Supporters of Catholic hospitals maintained that spiritual and temporal tending were equally significant in terms of treatment. They warned that public hospitals did not offer complete care because "the care of the soul is not the order of the day."[53] Catholic hospitals, by contrast, offered more effective treatment because, in addition to medical treatment, they offered "the sick in soul the blessing of a spiritual retreat."[54]

The hierarchy recognized that hospitals, like schools, could nurture and maintain a Catholic culture in New York. The *Freeman's Journal* in 1856 explained that hospitals offered the chance for "the erring child of the Church" to return "to his God."[55] At an inaugural banquet for the Long Island College Hospital in 1858, Hughes was chided by another speaker who claimed that while "the church would generate a thousand hospitals . . . we never heard of an hospital generating a church."[56] Hughes and his supporters disagreed. They hoped the "Sister Magdalens" could do just that.

Hughes and his successors actually had very little to do with the opening of the first Catholic hospitals in New York and Brooklyn. Without exception, these hospitals were founded by sisters. But through their joint efforts, inside public institutions and in building Catholic ones, sisters and priests—along with significant lay support—made the Roman Catholic Church a major participant in New York's growing hospital system, which included both city-run and privately run hospitals. By the century's end the church was active in the religious affairs of the municipal hospitals and on equal footing with the Protestant clergy. Bellevue Hospital's first Catholic chapel, completed in 1888 with private funding, was a very visible testament to a Catholic presence there and in other public institutions. At the same time, the church was also very much involved in the development of private hospital care.[57]

While all this Catholic activity and success can be considered as part of the church's growth and power in New York City, the two efforts had differing goals. The origins of Catholic hospitals were rooted in nineteenth-century economic and social circumstances, but the hospitals were also decidedly the result of the leadership and involvement of women within the church, the sisters. All of them came to be and thrived because of women like the Sisters of Charity at St. Vincent's. When we consider their involvement, the origins of Catholic hospitals become clearer and certainly less inevitable.

Sisters' involvement in health care was not simply a female accompaniment to the hierarchy's concerns over what was going on in city hospitals. Sisters' efforts in hospital development derived from their own view of church and society and their role in both. Their hospital work was based on the cornerstones of their own lives, which were community, service, and spirituality. Temporal concerns about territory and power—very much the focus of the hierarchy's push for a Catholic presence in municipal institutions—had little relevance. Although nineteenth-century separate sphere ideology was a factor in the story, it was more than a gendered division of labor that brought the hierarchy into the world of New York politics and government, and the sisters into Catholic hospitals. Sisters offered alternative models for both public Catholicism and institutional health care. Because the sisters were unique, their hospitals were too.

## CHAPTER TWO

# "To Serve Both God and Man"
# *The Sisters*

Prosaically celebrated in Catholic literature in lyrical terms as "creature[s] vowed to serve both God and man" who could "heal the body and save the soul," sisters were very hard workers.[1] The Sisters of Charity who organized St. Vincent's Hospital in 1849 were the first of six communities to open hospitals in New York City in the nineteenth and early twentieth centuries. After the Civil War, they opened the New York Foundling Hospital in 1869; that would expand to include a maternity hospital, St. Anne's (1880), and a children's hospital, St. John's (1881). Seton Hospital for tuberculosis patients was established in 1894; another, St. Lawrence, in 1915. All these were in Manhattan. In Brooklyn, the sisters' earliest efforts in hospital development were at St. Mary's Female Hospital (1868). St. Mary's was the foundation for two later hospitals: St. Mary's General, which first accepted patients in 1882, and Holy Family, organized as a general hospital in 1909. On Staten Island they founded a general hospital, also called St. Vincent's, in 1903.

Other communities opened hospitals as well. The Franciscan Sisters of the Poor opened St. Peter's Hospital in Brooklyn in 1864; St. Francis' and St. Joseph's in Manhattan, both general hospitals, in 1865 and 1882, respectively; and St. Anthony's, a tuberculosis hospital in Woodhaven, Queens, in 1915. Other Franciscans took over St. Elizabeth's Hospital in Manhattan in 1891. The Misericordia Sisters founded Misericordia Hospital, originally organized as the New York Mothers Home of the Sisters Misericorde in 1888. Missionary Sisters of the Sacred Heart founded Columbus Hospital, later called Cabrini, in 1892. Dominican Sisters and Sisters of St. Joseph each organized two general hospitals in the Diocese of Brooklyn (which included Brooklyn and Queens) between 1869 and 1910: St. Catherine's, St. John's, St. Joseph's, and Mary Immaculate.[2]

The motivation for these sisters' involvement in hospital development might appear self-evident. The caretaking quality of the hospital work was very much in keeping with their mission of service, and the domestic nature of nursing was appropriate to their status within the church. Hospital work was all very "sisterly."

This simple explanation is only partially true, and it supports inaccurate stereotypes about nineteenth-century nuns, specifically that sisters' lives have been static over several centuries and that sisters have been passive participants in their own lives and work. When Catholic hospitals opened, the Catholic press was fond of placing them in a long church tradition, suggesting that the sisters' hospital work reestablished a rightful position for the church. *Catholic World,* for example, noted that nursing was a Christian tradition upset by the Reformation, which by "sweeping away the work of the pious ages" had taken it from the hands of the religious orders where it belonged.[3] But the sweeping was long over and religious life for women much different than in the "pious ages," even among those who included some nursing in the monastic life of prayer and separation from the rest of the world.

American sisters' decisions to open hospitals in the nineteenth century represented a new mission among sisters rather than a resumption of duties. Most of the communities that organized hospitals were relatively young, founded only in the early nineteenth century, and had no particular historical predisposition to either nursing or hospital work. Moreover, when the first Roman Catholic sisters began working in the United States, in the very late eighteenth and early nineteenth centuries, most were teachers. As Barbara Misner showed in her research on the first American sisterhoods, those who did nurse, approximately one-quarter, did so along with other work, and it was never their main focus. Initially, they simply visited sick people, but over time they would become known for the quality of physical care—the nursing—they offered on these visits.[4]

Fearing that native-born Americans would not easily accept the idea of nursing sisters, the church hierarchy initially recommended that the Sisters of Charity stay away from nursing. Writing to the community's founder, Elizabeth Seton, in 1811, the first American bishop, John Carroll, expressed his opinion that a "century at least will pass before the exigencies and habits of this Country will require and hardly admit of the charitable exercises toward the sick, sufficient to employ any number of the sisters out of our largest cities." He concluded that instead, the sisters ought to consider education as the "permanent object of their religious duty."[5]

Carroll's advice was based on early-nineteenth-century realities about the Catholic Church in the United States. When Carroll wrote to Seton, Catholicism in the United States was still close to its colonial roots: Catholics were geographically dispersed throughout the country but mainly lived in the upper South, and their numbers were small in comparison to the Protestant population. In contrast, by the middle of the century, the American Catholic Church was overwhelmingly immigrant, clustered in cities, and increasing in such numbers that it seemed to many to challenge Protestant hegemony. It was in this context that sisters began to open hospitals.[6]

All of the communities that opened hospitals in New York and Brooklyn were, to different extents, part of a revolution in religious life: they were very different from the monastic orders of the Middle Ages. American hospital sisters were part of a movement initiated by seventeenth-century European Catholic women which redefined how sisters lived. Church leaders in post-Reformation Europe sought to restrict the lives of women religious, but women organized numerous new sisterhoods, which organized their lives and work outside of a traditional cloistered model. These sisterhoods, eventually referred to as active communities, became increasingly more popular than the older model of convent life, that of contemplative sisters. Active communities organize their lives differently than contemplative sisters do; prayer remains fundamental, but their life is also a mission of work beyond what it takes to maintain their convent.[7]

The change began in France in the seventeenth century when Roman Catholic religious women organized their lives and work among—not apart from—other Catholics and, in particular, those who were poor, sick, and needy. The movement grew and flourished in Catholic Europe in the late eighteenth and early nineteenth centuries. These communities would eventually send women to the United States, and many of those emigrants, along with American-founded sisterhoods organized on this new model, would include hospital work in their mission.[8]

While the commitment of these communities was service to the poor, which sometimes included care of the sick, there was never any mandate that sisters get into the business of opening hospitals. None of the communities of sisters that opened hospitals in New York was specifically organized either to nurse or to open hospitals. 'Sisters' involvement in hospital development was not an inevitable extension of their charitable mission; it was a deliberate choice made by sisters and their supporters who believed that their goals, which included both the alleviation of suffering and guidance on a path to salvation, would be well served in health care. In New York, sisters saw a city awash with suffering and imagined a sanctuary for the people they had devoted their lives to helping—immigrants and their children—available only in institutions at best inhospitable, at worst frightening. Moreover, they saw a link between sickness, the health care services available in New York, and the moral and spiritual lives of immigrant New Yorkers.

Hospital work was just one of several ways sisters attempted to help immigrant New Yorkers deal with the hardships of life in New York during the nineteenth century. They also organized orphanages, foundling homes, what would come to be called nursing homes, and a variety of schools. Like other nineteenth-century benevolent women, New York sisters were especially concerned with the plight of other women. A number of their hospitals began as homes or refuges for mothers and their children.[9]

There were several instances in which sisters' work among the immigrants actually broke with tradition and the regulations of their religious community. In the most extreme example, some of the Sisters of Charity in New York separated from their Maryland motherhouse in order to pursue work that the rules of their community would not allow. In the mid 1840s, Bishop John Hughes asked the Sisters of Charity in New York to take charge of the entire Roman Catholic Orphan Asylum where, at the time, they were responsible only for girls. Initially the Christian Brothers, and later a group of laymen, had charge of the boys, but neither was able to successfully manage that half of the institution. The sisters' superiors in Maryland instructed the New Yorkers to turn down Hughes's request because they interpreted the rules of their community to mean that Sisters of Charity could not care for boys. Hughes, who anticipated that he would need sisters who could teach both girls and boys in the parochial system he hoped to build, encouraged the New Yorkers to establish a separate congregation in the New York Archdiocese. Each sister in New York decided for herself whether to remain there or return to Maryland; of the sixty-two sisters in New York at the time, thirty-three remained.[10]

Other sisters were able to interpret the general regulations of their community broadly and met with little, or only delayed, resistance from a distant motherhouse. The Sisters of St. Dominic were officially a cloistered order until 1896, but they engaged in seemingly noncloistered hospital work by extending the confines of their cloister to include St. Catherine's Hospital in Brooklyn. (This was not unlike the model of monastic hospitals referenced in promotional literature like the *Catholic World* article noted earlier.) They also nursed at a public institution, Brooklyn's Hospital for Contagious Diseases, during a smallpox epidemic in 1893.[11]

While dedicated to their work with the immigrant poor in the secular world, sisters remained equally committed to their own religious life. Each community lived according to a particular "Rule" that articulated its philosophy and objectives, and guided the sisters through a daily routine. Sisters founded and ran hospitals without any master plan of hospital care to follow, but the way they went about their work adhered to this Rule upon which their lives were based.[12]

All rules were similar in that they included spiritual directives, ritual, and practical information, but not all were identical and each originated in a particular historical and cultural context. Some had European origins. The Sisters of St. Joseph, for example, who founded St. John's and St. Joseph's hospitals, came from a community founded in France in the mid-seventeenth century. It was organized specifically to do charity work outside the confines of a convent and was not cloistered. Their Rule was very flexible about the work they were allowed to do and was one of the factors influencing their move to the United States in 1836. The laywoman who financed the trip of six Sisters of St. Joseph from France to St. Louis in 1836 chose them because their Rule was "not so restricted as in some other Orders."[13]

The Sisters of Charity, the first community of women religious founded in the United States, was organized around the Rule of the French Daughters of Charity. Like the Sisters of St. Joseph, the Daughters of Charity were never a cloistered order. In 1810, under the direction of Elizabeth Seton, their Rule was translated and adapted for the new American congregation. It outlined the Sisters of Charity's mission as follows: "The Principal end for which God has called and assembled the Sisters of Charity, is to honor Jesus Christ, . . . by rendering him every temporal and spiritual service in their power, in the persons of the poor, either sick, prisoners, insane or those who through shame would conceal their necessities."[14] Their Rule also included details about caring for the sick that combined religious concerns with practical matters of nursing. It instructed sisters to encourage patients "to make a good General Confession, or to die well" and cautioned them that while remaining compassionate, they were "never to give any nourishment or medicine contrary to orders."[15]

The sisters' religious faith was a tremendous source of strength amid adversity. As Margaret Thompson has noted, "Only faith can explain the perseverance, serenity and strength with which some of the crises sisters faced were endured and overcome."[16] More importantly, the sisters' desire to achieve a state of perfect grace propelled them into their work with the poor and sick. Their spirituality was the focus of their lives, the catalyst for their work, and a significant factor creating and defining their workplace.

In the sisters' eyes, the poor and needy were their fellow sufferers in the mystical body of Christ, the church. As such, they sought to be among them—and like them—in many ways. The statutes of the Franciscan Sisters who opened St. Peter's Hospital in 1864 were particularly concerned with this identification and specified that the "Congregation is determined to possess no property beyond what is indispensable for the convent and chapel and the maintenance of the Congregation; as they work and care for only the poor, so they will also live with the poor and like the poor from donations of charity."[17] Other communities had less stringent rules about what they might own as a community, but all were dedicated to service.

There was sometimes concern among sisters that secular activity, what they referred to as the external life, and being within the secular world could compromise the religious life. Franciscan superior Frances Shervier, for example, worried about the effect that living in New York City might have on her sisters. The German community came to the east coast of the United States via Cincinnati, their first mission in the United States. When asked to send sisters to New York, Shervier wrote from Germany to the American superior in Ohio about her concerns. "In regard to the foundation of the two houses in the suburbs of New York, I think that it would be a good thing; still I have a fear that external activity in the vicinity of this city, so devoted to pleasure, must be extremely dangerous." She had heard "the worst reports about this city." Yet Shervier put her fears aside, writing her "dear Sisters" that if they "sufficiently consider all this and still think that we may venture to make a foundation there for visiting nursing, I will willingly consent. Yes, I must admit, I should like *very much* to have a house in New York or one of its suburbs." Following this correspondence, Franciscan sisters moved east. Their first work in the New York area was in Hoboken, New Jersey, where they opened St. Mary's Hospital in 1863; from that foundation, their work spread to Brooklyn and Manhattan. Emphasizing their commitment to a spiritual life, the Franciscans maintained the European practice of supporting recluse sisters, cloistered members of the congregation who devoted their entire life to prayer.[18]

But work and spiritual pursuits were ultimately one and the same for hospital sisters. Regulations of the Sisters of Charity directed that sisters were to pay close attention to the manner in which they carried out their work in the outside world. It was not only important to accomplish a task but to do it within the context of a particular kind of life: "The exercises of an interior and spiritual life must accompany . . . exterior employments."[19]

The Sisters of Charity were the first indigenous American community of women religious but not the first Roman Catholic sisters in what would become the United States. European sisters were in North America during the colonial period. The earliest were French Ursulines who arrived in New Orleans in 1727. Carmelite nuns founded the first religious community in the United States in Port Tobacco, Maryland, in 1790. Other European groups (Carmelites, Poor Clares, Visitation Nuns) also sent out small missionary bands in the late eighteenth and early nineteenth centuries.[20]

More spectacular growth in the number of convents and sisters occurred between 1830 and 1850 as American orders were founded and other European congregations established American foundations. Estimates indicate that there were approximately 1,500 Roman Catholic sisters in the United States in 1850. Some acquired a reputation for skilled nursing even before they began opening hospitals. In particular, the Sisters of Charity were lauded for their work in the cholera epidemics of 1832 and 1849.[21]

In the years prior to the Civil War, however, there were less favorable images of and attitudes about Catholic sisters in Protestant America. Sisters featured prominently in nativist attacks on the Catholic Church in the early nineteenth century; the most infamous instance was when an angry mob burned an Ursuline convent in Charlestown, Massachusetts, in 1834. In the antebellum years perhaps the best known Catholic nun in America was Maria Monk, a woman who claimed to have been impregnated by a priest while held captive in a convent from which she fled to avoid having her baby murdered. *The Awful Disclosures of Maria Monk,* first published in 1836, was one of several popular fictional accounts of convent life publicized as ex-nun autobiographies and concocted to expose the villainy and immorality nativists proclaimed to be found within convent walls. Throughout the 1830s and 1840s, so-called convent reformers lobbied for state regulation and inspection of convents, and, with the impetus of the nativist Know-Nothing movement in the 1850s, the Massachusetts and Maryland legislatures both established committees to investigate convents.[22]

These reactions grew out of general nativist views of Roman Catholics but also reflected specific attitudes about sisters. Like other Catholics, sisters were accused of papal attachments, which supposedly rendered them incapable of loyalty to America. Their opponents also charged that sisters proselytized whenever possible, something they felt was particularly dangerous as many were teachers. As historian Joseph Mannard has suggested, the particular lifestyle of sisters evoked hostility as well. Their convent life was characterized as an aberrant and dangerous rejection of the domestic ideal.[23]

At the outbreak of the Civil War some Catholics saw an opportunity to shape a new public attitude about sisters and, by extension, about the Catholic Church in general. As one priest noted, "A little band of devoted Sisters, ministering like angels amidst the soldiery, will do away with prejudices; and show the beauty and resources of the Catholic faith . . . much more forcibly than all volumes of arguments and evidences."[24]

In New York City, however, Archbishop Hughes was reluctant to see the sisters in his diocese leave for war. Writing to the bishop of Baltimore in May 1861, he expressed his concern about the possible involvement of New York sisters: "There is also another question growing up, and it is about nurses for the sick and wounded. Our Sisters of Mercy have volunteered . . . I have signified to them, not harshly, that they had better mind their own affairs until their services are needed. I am now informed indirectly, that the Sisters of Charity in this diocese would be willing to volunteer a force of from fifty to one hundred nurses. To this proposition I have strong objections. They have as much on hand as they can accomplish."[25] While Hughes worried about what would happen to Catholic institutions in New York if sisters left, other communities in the United States went to war. The War Department, desperate for nurses and racked with internal controversy over how military nursing for the Union Army should be organized, was usually glad to have them; sisters also nursed in Confederate military hospitals.[26]

Despite Hughes's objections, twenty-seven New York sisters ultimately worked in U.S. Army hospitals. In June 1862, Secretary of War Edwin Stanton asked the Sisters of Mercy to take charge of a military hospital in North Carolina, and they did. They remained until April 1863, accompanying the hospital as it moved across the battlefields of North Carolina. Sisters often suffered along with the patients they nursed, and some sisters became sick themselves, but unlike a sister in a Maryland hospital who died while attending wounded soldiers, all of the New Yorkers returned home.[27]

The Sisters of Charity remained in New York but contributed to the war effort too. At the request of the War Department, they ran a military hospital, St. Joseph's, and stayed there until it closed in 1866. Primarily for the care of amputees and crippled soldiers, it was located near the community's original motherhouse in New York at McGowan's Pass at 109th Street and Fifth Avenue.[28] Still, not all New Yorkers welcomed the sisters' participation. The War Department also considered placing the Sisters of Charity at another military hospital on Lexington Avenue and 51st Street but did not because they thought that Protestant groups involved there would object.[29]

There was opposition in the Union Army as well. Although the Surgeon General would write that "No one can bear fuller or more willing testimony to the capability and devotion of Sisters of Charity than myself,"[30] Dorothea Dix, superintendent of women nurses for the Union Army, had other thoughts. She claimed that sisters would upset the hospital chain of command because of a primary allegiance to their own superior. Dix's views on nursing mirrored those of the English reformer Florence Nightingale, who had made a name for herself and her nurses in the Crimean War. Like Nightingale, Dix considered the treatment of illness, even battle wounds, in terms of both moral and physical processes of healing. Both women believed that female nurses, not male physicians, should hold central authority in hospitals because they were best suited to bring moral authority into patient care. (Dix's objections to sister nurses were identical to those raised in England when Nightingale included sisters in her Crimean nursing staff.)[31]

In her quest to become a nurse in years when nursing was not considered at all appropriate for women of her class, Nightingale thought she might find an answer to her dilemma in the Catholic Church. Before her work in British army hospitals, she was attracted to Catholicism, in part because of her admiration for Catholic sisters and the work they did. Although her thoughts of conversion eventually disappeared, her interest in sisters as nurses remained. Most histories of nursing mention Nightingale's visit to Kaiserworth, a Protestant community of women in Germany whose members did not take final vows, but she also made an effort to train with Catholic nursing sisters. In 1853 she entered a convent in Paris to train with the Sisters of Charity. She became ill and left almost immediately, but ten sisters (from another community) accompanied her to the Crimea in 1854.[32]

Dix's biographer David Gollaher concludes that a personal anti-Catholicism was the real root of her objections to nursing sisters. While that may well have been the case, she was not entirely wrong: sisters ultimately obeyed a higher authority and had their own priorities, which included their spiritual lives. Even in the midst of a battlefield hospital, they would strive to maintain their unique lifestyle: their Rule traveled with them. Once a community of sisters identified this work as part of its mission, however, there would be little chance that they would cause the kind of revolt against hospital authority Dix imagined. In fact, just the opposite was true.[33]

Favorable opinions about nursing sisters in the Civil War demonstrate that other volunteers did not bring the kinds of skills to their work that sisters did. One Civil War physician noted that sisters "adapted themselves admirably to their new duties" without fussing about their accommodations like previous nurses. Notably, the sisters he referred to with such lavish praise were a teaching order. Like most other Civil War nursing volunteers, they had no prior formal nursing experience. In his opinion, their selflessness was their greatest nursing skill. He recognized that the lifestyle of the sister embodied what Florence Nightingale had identified as characteristics necessary for an effective hospital nurse. The Catholic sister was selfless, chaste, obedient, and very devoted to her work. She also took orders well and had no fears about what she ought to be doing. Sisters were trained to endure, and they did so amid the sounds, smells, and realities of battlefield medicine.[34]

The American public's image of the Catholic sister changed dramatically by the end of the Civil War because of the reputation sisters earned as nurses in military hospitals. As Mary Ewens notes in her study of American sisters in the nineteenth century, "The Civil War years brought countless stories of their work as 'Angels of Mercy' . . . The sister, who had been the object of hatred, insult and persecution, suddenly became the subject of highest praise."[35]

Sisters' reputation, particularly with regard to their devotion to their work, continued to make them popular nurses among non-Catholics in New York during times of crisis. Health officials in New York and Brooklyn asked sisters to work in public hospitals during two public health emergencies. While their skill as nurses was part of the reason their help was requested (teaching sisters were not asked in these instances), another factor was that no one else was anxious to be exposed to contagious diseases. In 1866, five Sisters of Charity went to Ward's Island to nurse at the cholera hospital. In 1893, in the midst of a smallpox epidemic, the Brooklyn Board of Health asked the Dominican Sisters from St. Catherine's to send sisters to the City Hospital for Contagious Diseases.[36]

Because this dedication was difficult to duplicate, one private non-

| No. | Date | Name | Residence | Age | Civil Condition | Occupation | Religion | Nativity | How Long in U.S. | How Long in Borough | Diagnosis | Discharged | Results | Died |
|---|---|---|---|---|---|---|---|---|---|---|---|---|---|---|
| 4869 | Jan 1 1904 | Jos. Laughlin | Astoria L.I.C. | 74 | W. | None | Cath | [Ire]land | 50 yrs | 50 yrs | Chronic Bronchitis | Jan. 1. 1905 | Cured | |
| 4870 | 1 | Jos. Foy | 59 7th St. L.I.C. | 50 | M. | Laborer | Cath | U.S. | Life | Life | Paralysis Agitans & Chronic Rheum. | Jan 5. 1904 | Improved | |
| 4871 | 1 | William O'Brien | 374 Oakland Av. | 31 | S. | Laborer | Cath | U.S. | Life | Life | Rheumatism | March 16. 1904 | Cured | |
| 4872 | 1 | Adam Brown | 115 Greenpoint Av. | 40 | S. | Laborer | Cath | U.S. | Life | Life | Bronchitis | Jan 2. 1904 | Cured | |
| 4873 | 1 | Michael Duffy | 363 Jackson ave. | 48 | S. | Driver | Cath | [Ire]land | 47 yrs | 47 yrs | Bronchitis | Jan. 9. 1904 | Cured | |
| 4874 | 1 | John Tuckstil | Newtown L.I. | 38 | S. | None | Cath | U.S. | Life | Life | Cardiac Syncope | | | March 18. 1904 - [illegible] |
| 4875 | 1 | Patrick Hussion | 115 Bowbridge St. | 33 | S. | Laborer | Cath | U.S. | Life | Life | Acute exacerbation, chronic Nephritis | Jan 3. 1904 | Cured | |
| 4876 | 1 | Minnie Smith | 186 Vernon ave. | 7 | S. | School girl | Cath | U.S. | Life | Life | Pneumonia | Jan. 24. 1904 | Cured | |
| 4878 | 1 | John Hughes | Astoria L.I.C. | 69 | M. | Laborer | Cath | [Ire]land | 40 yrs | 40 yrs | Rheumatism | May 12. 1904 | Cured | |
| 4879 | 1 | Nicholas Rooke | 173 Jackson ave. | 43 | S. | Laborer | Cath | U.S. | Life | Life | Rheumatism | Jan. 21. 1904 | Cured | |
| 4882 | 1 | Annie Smith | 6th St. L.I.C. | 8 | S. | School girl | Cath | U.S. | Life | Life | Pneumonia | Jan. 24. 1904 | Cured | |
| 4883 | 1 | Carl Fastle | 241 Franklin St. | 60 | M. | Laborer | Cath | [Ger]many | 37 yrs | 37 yrs | Endo-Carditis | | | Jan. 1. 1904 - [illegible] |
| 4956 | 1 | Joseph Gunne | 157 West Ave. | 34 | S. | Laborer | Cath | U.S. | Life | Life | Rheumatism | Jan. 23. 1904 | Cured | |
| 4930 | 1 | John Rower | 139 Jackson ave. | 48 | S. | Laborer | Cath | [Ire]land | 28 yrs | 28 yrs | Bronchitis & Articular Rheu. | Jan 2. 1904 | Cured | |
| 4962 | 1 | Jos. Steedman | Manhattan ave. | 26 | S. | Driver | Pro[t] | [illegible] | Life | Life | Rheumatism | March 20. 1904 | Cured | |
| 4967 | 1 | Mary Donahue | 64 Wilbur Ave. | 57 | M. | Housework | Cath | [Ire]land | 20 yrs | 20 yrs | Acute articular rheumatism | Feb. 1. 1904 | Cured | |
| 4970 | 1 | John Ducats | North Beach | 23 | S. | Driver | Prot | U.S. | Life | Life | Nephritis & Rheumatism | Jan. 18. 1904 | Cured | |
| 4971 | 1 | Ellen Evers | Astoria L.I.C. | 45 | M. | Domestic | Cath | U.S. | Life | Life | Articular Rheumatism | April 21. 1904 | Cured | |
| 4973 | 1 | Jos. Maxwelle | Greenpoint | 35 | M. | Laborer | Cath | U.S. | Life | Life | Bronchitis & Rheumatism | March 12. 1904 | Cured | |
| 4975 | 1 | Charles Bertina | Steinway Ave | 43 | M. | Laborer | Cath | [illegible] | Life | Life | Pneumonia | May 20. 1904 | Cured | |
| 4976 | 1 | Mary Dowd | 98 Hunter Ave | 36 | M. | Domestic | Cath | U.S. | Life | Life | Chronic Articular Rheumatism | March 19. 1904 | Improved | |
| 4978 | 1 | Kate Welsh | Van Alst Ave. | 65 | S. | Housework | Cath | [Ire]land | 30 yrs | 30 yrs | Chronic Rheumatism | Jan. 24. 1904 | Improved | |
| 4979 | 1 | John Hannigan | Greenpoint | 35 | S. | Driver | Cath | U.S. | Life | Life | Bronchitis | March 14. 1904 | Cured | |
| 4981 | 1 | Jos. Burke | 6th St. L.I.C. | 40 | S. | Laborer | Cath | [Ire]land | 20 yrs | 20 yrs | Rheumatism | May 17. 1904 | Cured | |
| 4992 | 1 | Rose Strasser | Arverne L.I. | 18 | S. | Domestic | Cath | U.S. | Life | Life | [illegible] poisoning | Jan 21. 1904 | Cured | |
| 4993 | 1 | Lizzie O'Rourke | 279 Jackson ave. | 32 | M. | Housework | Cath | U.S. | Life | Life | Malaria | Jan 9. 1904 | Cured | |
| 5060 | 1 | Jos. McCoy | Concord St. | [illegible] | | Laborer | Cath | U.S. | Life | Life | Rheumatism | Feb. 11. 1904 | Cured | |
| 5059 | 1 | Thomas Brady | 23 Greenpoint Av. | 33 | S. | Laborer | Cath | U.S. | Life | Life | Heart Disease | June 1. 1904 | Cured | |
| 5152 | 1 | Jos. Lynch | 643 Steinway ave. | 33 | S. | Laborer | Cath | U.S. | Life | Life | Articular Rheumatism | May 26. 1904 | Cured | |
| 5157 | 1 | John Reynolds | Nassau Ave | 52 | M. | Driver | [illegible] | [illegible] | Life | Life | Tonsilitis | Feb. 11. 1904 | Cured | |
| 5123 | 1 | Aug. Fisher | 747 8th Ave. | 47 | M. | Laborer | Prot | [Ger]many | 20 yrs | 20 yrs | Gas-inhalation | | | Jan. 3. 1904 - [illegible] |
| 5125 | 1 | Richard Trimey | 175 India St. | 45 | M. | Laborer | Cath | [Ire]land | 14 yrs | 14 yrs | Pneumonia | | | Jan. 2. 1904 - [illegible] |

Catholic hospital in New York City, the French Hospital, recruited a community of sisters to work. In 1885, the French Benevolent Association decided to hire the Marianite Sisters of the Holy Cross to nurse and supervise at its hospital because the sisters were reliable. Members of the association decided that "even though our means were better, we would always have to get our personnel among persons having nothing to do for the time being and ready to give up at any time a task which they had accepted because there was nothing else at the time."[37]

Despite praise, individual sisters remained anonymous figures in public contemporary accounts of the city's Catholic hospitals. Acknowledgement of their special place in charity work because of their "supereminent devotion" is characteristic of the way their hospital work was described.[38] While many believed that a sister's charity was "pure and unselfish"—all to the betterment of the institution she served—in most cases she remained in the background and the actual details of her activities and life within the hospital were left unrecorded.[39]

Hospital annual reports rarely included a sister's name in pages filled with lists of benefactors, physicians, and church officials, but sisters made financial contributions too. The communities of sisters that founded hospitals supported them in a number of ways, and would go to great lengths to do so. With few other resources, the sisters at St. Francis' in Manhattan begged for money and food. When Mary Immaculate Hospital opened, sisters there depended on monthly advancements from their Dominican Motherhouse in Brooklyn to make up the difference between expenditures and receipts. They also went out soliciting donations.[40]

When a community of sisters operated a number of different institutions, it could and did arrange loans between them. While this was usually transacted formally with contracts and the loans carried interest, sisters went about all these obligations in a sisterly sort of way. The comments of a Sister of Charity at St. Lawrence Hospital, for example, who reminded her colleague at St. Joseph's Home for the Aged of a debt due St. Lawrence were not those of a typical banker. "I hope somebody will leave you a half of a million dollars. Anyway, may the October Angels bring you what is better than gold,—graces in abundance." When a community could not finance its sisters, it borrowed money for them. The Sisters of Charity assumed responsibility for a loan that got St. Vincent's started by mortgaging their real estate.[41]

*Fig. 4. (opposite) Record Book St. John's Hospital,* June 7, 1891–December 31, 1894. Sisters of Saint Joseph, Brentwood, New York

Sisters contributed to their own anonymity. In the spirit of humility they were not anxious for personal recognition, but within their religious communities they carefully preserved mementos and documents that recorded their hospital work and achievements. An annual report kept for a community's records, for example, has the names of the sisters neatly handwritten on the back cover. Notes kept by a continuous line of convent annalists reveal diverse personalities among sisters, even when autobiographical information is almost exclusively imparted in glowing terms. In a collection kept by the Sisters of Charity, one hospital administrator is characterized for her strong spirituality, another for her practicality and efficiency; these remarks suggest a distinction between the two women's administrations.[42]

There were differences among religious communities of women as well. Each of the communities that founded hospitals had strong roots among one immigrant group in New York and, as a result, they differed culturally from each other. With the exception of the Sisters of Charity, which was founded by the American-born Elizabeth Seton, the first members of the other New York communities in hospital work were immigrants themselves. The Dominicans and Franciscans were communities from Germany that sent several members to the United States in the 1850s; the Misericordia Sisters came to New York from Québec; and the Missionary Sisters of the Sacred Heart were founded in Italy by Mother Frances Cabrini, who came to New York with a small group of sisters in 1889.[43]

Like other immigrants, sisters, especially in the antebellum period, often faced uncertainties when they arrived in New York, even with careful planning. The first of the Dominican sisters to come to New York in 1853 expected to be met on their arrival by a priest from western Pennsylvania, in whose parish they planned to teach. According to convent records, when the priest did not show, the sisters, who spoke no English, made their way to Manhattan's German parish, Most Holy Redeemer on Third Street, because they had a letter of introduction to the priests there. Through connections the priests at Holy Redeemer had with a priest in Brooklyn, the Dominicans soon settled into another German immigrant neighborhood in Williamsburg, Brooklyn.[44]

Sisters' lives in the United States differed from those of sisters in Europe. Some European convents enjoyed perpetual endowments or owned land that could support the sisters, but most American convents had no such financial backing when they began. A fledgling group in New York with limited support from a motherhouse either in Europe or the United States had to find ways to raise money before opening a

*Fig. 5.* A private room at St. John's Hospital in the early twentieth century. Hospitals hoped to attract paying patients with comfortable private rooms. While annual reports like the one in which this was published emphasized that hospitals were not exclusively for the treatment of Catholic patients, this photograph sent a slightly different message. Sisters of Saint Joseph, Brentwood, New York

hospital. In what would be a pattern for many communities, the Sisters of Charity opened a school for paying students in order to support themselves when they first arrived in New York.[45]

The Sisters of Charity had been founded in Maryland, but they had strong ties to one immigrant community in New York, too. Like New

York's Catholic hierarchy after 1842, a Sister of Charity was most likely to be of Irish descent. Indeed, one of the sisters who opened St. Vincent's Hospital was New York bishop John Hughes's Irish-born sister, Ellen Hughes, or Sister Angela Hughes as she was known in her religious community.[46]

Ethnicity was a strong factor in determining which community a woman would choose to join. While the number of communities of sisters in the United States grew dramatically in this period, most women chose a group they knew something about and, in most cases, one whose members shared their same cultural background. The roster of the sisters who worked at St. Vincent's throughout the nineteenth century consists overwhelmingly of women with Irish surnames. The first U.S. entrants to the Franciscan congregation, Julia Kayller and Susannah Oeschsner, were both German-born. So were all the administrators at their St. Peter's Hospital in Brooklyn between 1864 and 1912. Similarly, Dominican novices in the same period were almost entirely from German families, either German-born themselves or the daughters of German parents. Of the five Irish names appearing on the Dominican community's roster prior to 1901, at least two were probably more culturally German than Irish—Agnes Sheridan and Mary Ryan (later Sister Radegundis and Sister Charles Edward) had both been raised by the Dominicans in convent orphan asylums. For this community, cultural bonds were particularly important since they remained predominantly German speaking until the early twentieth century.[47]

Catholic literature celebrated the idea of a wealthy upper-class woman who abandoned the luxuries of the world for the convent in sentimental piety, as in the poem "The Sister of Charity" by Gerald Griffin:

> She once was a lady of honor and wealth.
> Bright glow'd in her features the rose of health
> Her vestures were blended of silk and of fold.
> And her motion shook of perfume from every fold;
> Joy revell'd around her—love shown at her side—
> And gay was her smile as the glance of a bride,
> And light was her step in the mirth sounding hall
> When she heard of the daughters of St. Vincent de Paul.[48]

Yet while a number of the earliest American sisters came from well-to-do families, it would not remain so. It was overwhelmingly women of immigrant backgrounds who swelled the ranks of American sisterhoods in the late nineteenth and early twentieth centuries.

Education and prior association with a particular order also influenced a woman's decision about which community to join. Like the Dominican sisters Radegundis and Charles Edward, a novice often had personal associations with a community before she entered its community. She may, for example, have been educated by the sisters. Ellen Hughes attended St. Joseph's Academy in Maryland, which was run by the Sisters of Charity. Her decision to join the Sisters of Charity seems to have been anticipated. When she and her sister enrolled as students in 1823, her brother John asked the Sisters to waive the usual amount of money a student brought with her "should one . . . of his sisters evince a desire to join the community." Four months after her arrival at St. Joseph's, Ellen asked to be admitted to the congregation.[49]

Although education was one way a woman became acquainted with a community of women religious, a Catholic education was never a requirement to join a convent. However, parish affiliations with a particular religious community could influence which order a woman might join; an interested woman without her own connections might approach her parish priest who would then direct her to the community he knew best. Sister Mary Loretto Donahue, for example, one of the founders of St. Mary's Hospital in Brooklyn, came to the Sisters of Charity in 1870 at the suggestion of Monsignor Taaffe, an influential priest in Brooklyn. A priest at Our Lady of Sorrows in Manhattan suggested that Mary Pinning, born in Schleswig-Holstein, join the Brooklyn Dominicans in 1869.[50]

Converts could also enter a convent. Mary Pinning, like a small but noteworthy number of sisters, was a convert. Among the Sisters of Charity the precedent had been set with their founder, Elizabeth Seton, who had been Episcopalian. Like Seton, Mary Jerome Ely, who managed St. Vincent's Hospital from 1855 to 1861, was a convert. The daughter of an Episcopalian mother and a Presbyterian father, she converted to Catholicism after having attended the St. Joseph's school in Baltimore.[51]

There were (and are) several steps to membership in a religious community. As postulants, women were prepared as candidates for a congregation's training school, called a novitiate. As novices, they received the habit of the community and were accepted as candidates for full membership. After an established period of training and preparation, novices professed temporary vows. After another designated time period, a sister, if accepted by the community, could, according to the statutes of that community, make perpetual or final vows or renew the temporary vows. A professed sister was one who had made vows of poverty, chastity, and obedience.[52]

Entrance age varied, but it is clear that postulants were not necessarily very young women. The first three women admitted to the Dominican community for hospital work (and who remained in the congregation) were each over twenty years old when they entered the novitiate in 1869: Philomena Dumoulin and Anne Marie Kerling were both twenty-one and Thecla Streble was twenty-four.[53]

Economic concerns could keep women from entering at a younger age. Madeline Reuger, one of the initial group of Dominican sisters at St. Catherine's Hospital, first asked to join the congregation when she was sixteen. The Dominican superior felt that her entrance would be a financial hardship for her widowed mother who had several younger children and told the girl to stay at home until her youngest brother was old enough to work. She did, and entered the novitiate five years later when she was twenty-two. Other youthful requests were also postponed. Ellen Hughes asked to join the Sisters of Charity when she was seventeen, but they advised her to wait until she was one year older.[54]

Even when such care was taken to ensure success, not all women who entered the convent were cut out for religious life, and not all postulants or novices remained in the convent. Women left by their own choice and also on the decision of the community. Two of the five women accepted by the Dominicans in 1869 never took their final vows, which was not at all unusual. Describing a group of novitiates making their first profession in 1883, Dominican superior Seraphima Staimer noted that when the group was first received in 1881 there had been sixteen of them, and two years later, twelve remained. "Four of them did not seem to have a true vocation. Two left on their own accord and two were dismissed by us."[55]

Unhappy novices could be disruptive. Staimer dryly recorded her experiences with one candidate who had already had several unsuccessful convent experiences before she presented herself to the Brooklyn Dominicans. "She was not candid and deceived me in her first interview. As she promised amendment I took pity on her and sent her to Amityville to the Novitiate. She was so sullen and discontented that her companions feared her." After this woman half-heartedly attempted suicide, Staimer's patience ran out, and "We hurriedly got her passage and sent her back to Europe, bag and baggage." Another woman was similarly packed up and sent off because she was "a very restless, headstrong person, discontented and quarrelsome." Notably, both of these women were from Germany, not from the German immigrant community in New York. Most likely neither they nor their families were as well known to the community members in New York as later American candidates were.[56]

The preponderance of cultural and community ties meant that a woman's life as a sister would include people, language, and traditions she knew from home. An exceptional example is the life of a Dominican sister, Mother Caritas. Her entire life—105 years—was spent in Williamsburg, Brooklyn. She was constantly surrounded by family, friends, and familiar faces inside and outside the convent. (Two of her own sisters also joined the Dominican congregation.) Most of her life, as young Elizabeth Harth and as a sister, was spent in the parish of Holy Trinity Church. She was born there in 1861 and attended the Dominican-run parish school as a child. When she entered the community in 1879, their novitiate was located at the Dominican's Holy Cross Convent at Holy Trinity, so she did not move far from home. After a brief assignment at another parish school, she returned to Holy Trinity parish as a teacher and was eventually principal there. Throughout the course of her long career, she held a number of different positions within the congregation and often continued to live at Holy Cross convent. She retired in 1943 and lived there until her death in 1966.[57]

Within their convents, sisters lived physically close together and considered themselves members of a family. In some communities, this family identification extended beyond convent walls, to other houses of the same community, and often efforts were made to maintain bonds among houses. In Germany and the United States, all Franciscan sisters participated in the same meditation each Thursday evening and Friday afternoon. Convents within a community communicated with each other about their goings on: the progress they were making in their work and difficulties they faced. New York Sisters of Charity, for example, wrote to others in Halifax in 1857 about a recent visit of the bishop to St. Vincent's Hospital in Manhattan noting that, "His grace is much pleased with our work." A picture postcard of St. Catherine's Hospital at the turn of the century brought greetings from Dominican sister Mercedes to Sister Polycarp at another mission wishing "A merry Christmas to you and all dear Sisters in the woods." When a Franciscan sister died, a notice of her death went out to all houses.[58]

A letter from Franciscan founder Frances Shervier to Franciscans in Germany during her visit to the United States in 1863 shows one way sisters maintained connections within the community, even across great distances. She wrote to her "Dear, dear Sisters," from Ohio that "On August 9 a new foundation is to be made and on that same day there will be an investiture . . . I desire now that the Sisters pray very much for the blessing of God on the new work. On the day—August 9—the Sisters shall have an extra coffee with some thing besides." All the Franciscan sisters participated in one house's celebration.[59]

All aspects of a sister's world were carefully regimented and. prayer was a primary focus of the day's activities. In the 1880s, the regulations of the Sisters of Charity outlined a day that began at 4:30 A.M. and was organized into periods to be spent in work, prayer or spiritual reading, and meals. After a daily recreation period between 7:00 P.M. and 8:10 P.M., and then meditation and chapel in the evening, it was "In bed and lights out" at 9:15.[60]

The Dominicans' schedule was similarly organized. Sisters rose at 4:00 A.M., at which private prayer and meditation was followed by Mass and then breakfast. The rest of the day was divided into periods for meals, prayers, meditation, and different kinds of work. After supper in the evening, sisters ended their day with prayers, meditations, and a period of recreation, which for most communities meant communal reading and sewing.[61]

While all sisters observed a daily schedule of prayers, this was only one part of their regulated lives—food, clothing, and sleeping arrangements were prescribed as well. Some regulations could, and would, be adjusted over time by community superiors. Franciscans were originally only allowed to eat meat on Tuesday, Thursday, and Sunday, and regulations specified that it be chopped up and added to vegetables and potatoes. In 1858, their guidelines were changed to note that each sister was to receive a slice for herself. In 1882, the community decided to allow meat daily during June, July, and August to "enable the Sisters to have more strength for work." Other detailed Franciscan directives pertained to bedding, burial arrangements, and seasonal devotions. Community regulations like these were viewed as tools to assist the sisters in their quest for spiritual fulfillment. The hardships they often endured through choice were intended to bring them to a fuller identification with the sufferings of Christ.[62]

Convent life was not always smooth. Even within the strict discipline of religious life, personalities emerged and clashed. One of the house superior's roles was to minimize conflict, and it was a position which required skill and diplomacy. When Frances Shervier wrote Sister Felicitas, a new house superior and one of the Franciscans who organized St. Peter's Hospital in Brooklyn in 1859, she acknowledged that the superior was not a role always eagerly sought: "Well can I think that you were much surprised and affected when you received the information that you should be for a time the cross bearer." Shervier also included advice about how to best deal with individual sisters and manage the group most effectively. "Toward Sister Augustin be particularly kind and forbearing, without however failing in your duty; this Sister has a very good will and, I do not doubt, would therefore also receive greater graces from God if she will only faithfully correspond. Do not spare Sister Dominica too much; do not let yourself be influenced by her seniority in religion and her manner of domineering a little." Finally, she reminded Sister Felicitas that she had a council, sisters who were to be consulted about important matters, and cautioned her to "never act without having consulted with both these Sisters."[63]

The chain of command was an important component of religious life. Writing to Sister Augustin later that year Shervier tried to smooth out some problems by reminding her who was in charge. "Try to be submissive and cordial toward Sister Felicitas. If nature rebels interiorly and complains, humble yourself." As she had reminded Felicitas of her position, she also told Augustin that while a councillor, she "should always endeavor to speak respectfully with the Superior."[64]

The decisions superiors made included where individual sisters would work. The Dominican sisters appear to be unique among New York hospital communities in that they selected particular novices for hospital nursing prior to the nurses training movement. When plans were made to open St. Catherine's Hospital, several women were accepted specifically for hospital work. These women were received into the congregation as lay sisters rather than what were referred to as choir sisters. Choir and lay sisters were distinguished by both their work and their vows. Lay sisters were assigned exclusively domestic and manual chores. By designating hospital sisters as lay sisters, the Dominicans mirrored contemporary attitudes about hospital nursing, since until the development of the training school movement it was considered servile work. Lay sisters followed differently prescribed prayers and wore different habits: choir sisters wore a white habit with a black veil and lay sisters a black habit with a white veil. These first Dominican hospital sisters were received into the congregation as lay sisters in 1869 at the same time that six other women entered as choir sisters. The Dominicans maintained a distinction between choir and lay sisters until 1896.[65]

Some Sisters of St. Joseph in the United States, the Brentwood group among them, also continued the European tradition of differ-

*Fig 6.* First ambulance at St. John's Hospital Long Island City, The background of this photo suggests a rural setting, but Long Island City was industrializing, and the hospital was founded to serve local laborers. Photograph ca. 1891; Sisters of St. Joseph, Brentwood, New York

entiating between choir and lay sisters. In her study of the Sisters of St. Joseph in the United States, historian Patricia Byrne finds that American sisters were uncomfortable with the distinction; she quotes one superior who wrote quite bluntly: "No one wants to be a 'lay sister.'" Still, the distinction remained into the twentieth century. A photograph of Sisters of St. Joseph at their St. John's Hospital in Queens at the turn of the century includes choir and lay sisters.[66]

For most of the century, contemporaries would commend sister nurses for their work and remark upon the difference between sisters and nurses in other hospitals. As nursing historian Susan Reverby describes, in the early period of hospital development in the United States, nursing "remained vague and linked to a variety of women's duties." Until the introduction and implementation of hospital training programs for nurses, hospital nursing ranked very low "within the hierarchy of paid labor," and the profession did not always attract other women as devoted to their work as the Catholic sisters were.[67] By the 1890s the image of nursing and the training of nurses had changed from what it had been when sisters in New York first opened hospitals. After the successful implementation of hospital-based nurses training programs, nursing became a more acceptable occupation for women. The training school movement sought to differentiate nursing from domestic work and to bring discipline and skill to the task of nursing. Nursing students in these programs were selected carefully and educated in a nursing school affiliated with a hospital. In 1873 the first three hospital training schools opened in the United States, one at Bellevue Hospital in New York. Throughout the 1880s, these schools gained growing acceptance.[68]

At the same time, some nursing reformers began to criticize the Catholic nursing orders. They discredited nursing sisters because the sisters had achieved a very favorable reputation without the benefit of a training school education. In a pamphlet it published on nursing reform, the State Charities Aid Association of New York, a private organization that reported to the state government on conditions in hospitals and other institutions, criticized nursing sisters because "the aim of the sisterhoods, Roman Catholic sisterhoods especially, is apt to be a divided one, their own spiritual progress, and the spiritual good of their patients, being supreme; Miss Nightingale says, 'hospital nursing is a jealous lover; it claims the whole heart.' "[69]

While the reformers made it clear that they considered Catholic nursing sisters to be insufficiently trained, in some ways the reformers found the sisters worthy of emulation. Qualities attributed to the nursing sisters, particularly devotion and obedience, were deemed necessary traits for all nurses. A training school prospectus from Bellevue even quoted the "holy founder of the order of the Sisters of Charity." Bellevue expected its students "to be religious women" and although it was not required "that they should belong to any given sect," they were advised to follow in the footsteps of the Sisters of Charity, whose founder St. Vincent de Paul told them, "Your convent must be the houses of the sick. Your cell, the chamber of suffering. Your chapel, the nearest church. Your cloister, the streets of the city or the wards of a hospital. The promise of obedience, your sole enclosure."[70]

Catholic hospital sisters opened their own hospital nursing schools where they trained both sisters and lay women. The first in New York was probably the Sisters of Charity's school at St. Mary's Hospital in Brooklyn, which opened in 1889. Some Catholic hospital schools initially trained only sisters. At St. Catherine's Hospital, the first class began in 1894 and consisted of nineteen sisters but the third class, which began in 1909, included laywomen as well. The first class at St. Vincent's in Manhattan in 1891 included seven sisters and eight laywomen.[71]

Although nineteenth-century hospital sisters were best known for their role as nurses, they also performed many other duties in Catholic hospitals: they were cooks, pharmacists, clerks, and laundresses, too. Sisters were also always the administrators, as well as members of the board of trustees. Of course, none of these sisters received salaries for their hospital work, a point they often noted when soliciting funds. Hospital costs did include living expenses for the sisters, but as the French Benevolent Society noted when they first considered bringing sisters into their hospital, sisters lived simply and frugally so those costs were low.[72]

With their efforts in hospital development, sisters linked Catholicism and health care in New York. By 1915, the date most often referred to as the benchmark of hospital development in the United States, the Roman Catholic Church in New York, with a total of seventeen hospitals founded by women religious, was clearly a major participant in the emergent world of hospital health care in the city.[73] A daily presence in almost every facet of hospital life, sisters shaped the character of Catholic hospitals. If there was anything uniquely Catholic about those institutions, it was the result of something they did.

CHAPTER THREE

# "Consoling Influences"

## *Care and Treatment*

The log the Sisters of Saint Joseph began when they opened their hospital in Long Island City in 1891 catalogued admissions chronologically, noting age, address, nativity, occupation, religion, illness, and then the death or discharge of each patient. Discharges were additionally noted as cured or improved, and the cures far outnumbered those merely improved. This was the stuff of nineteenth-century hospital annual reports, which were typically optimistic, constructed as they were to establish a reputation and raise money, but the neat and consistent hand of the scribe at St. John's also suggests something of the atmosphere sisters created in their hospitals, one of order, tidiness, and piety. These characteristics mark a stark contrast to the precarious and tumultuous world their predominantly working-class and immigrant patients inhabited.[1]

St. John's was founded when hospitalization for certain kinds of illnesses was becoming more commonplace, particularly among urban immigrants. For most of the nineteenth century, however, medical treatment was not at all synonymous with hospitalization. Even in instances of severe illness or accident, hospitals usually did not offer medical care. Patients received treatment in their own homes, physician's offices, and urban dispensaries, which provided advice and medicine (but not beds) to walk-in patients unable to afford a private physician's care. In the first three-quarters of the nineteenth century most patients at general hospitals suffered from chronic illnesses; general hospitals typically did not accept patients with what were considered to be contagious diseases. Furthermore, all hospital patients were identified as much by their social and economic condition as their physical one—hospitals were a place of last resort for sick people with no other resources.[2]

Two factors not inherently related to each other gave hospitals this reputation. First, there was a strong social stigma attached to hospitalization because some of the earliest hospitals in the United States began as public almshouses. (Bellevue, the first hospital in New York, was part of the city almshouse until 1849.) Second, the therapeutic techniques used for most of the century minimized the necessity of hospital-based treatment. What could be done for a patient medically could take place as easily outside a hospital. In the pre-Lister era, physicians could just as easily perform even the most extreme measures—bleeding and purging, even surgery—in a patient's home or the doctor's office. Other, less heroic, treatments like poultices, medicinal teas, and baths similarly required no special equipment or space.[3]

Still, the number of hospitals in the United States, especially in urban areas, increased dramatically in the second half of the century. Paralleling national trends, the number of Catholic hospitals increased in New York after the Civil War. Three Manhattan institutions, St. Vincent's Hospital, the New York Foundling Hospital and St. Francis' Hospital, and two in Brooklyn, St. Peter's Hospital and St. Mary's Female Hospital, opened before 1870. Between 1870 and 1900, sisters also ran the Mothers Home of the Sisters Misericorde, St. Elizabeth's, Seton, and Columbus Hospitals in Manhattan; St. Catherine's Hospital and St. Mary's General Hospital in Brooklyn; and St. John's Hospital in Long Island City. They opened five new hospitals in the first decade of the twentieth-century: St. Vincent's on Staten Island in 1903, Mary Immaculate in Queens in 1904, St. Joseph's in Far Rockaway in 1905, St. Lawrence in Manhattan in 1906, and Holy Family in Brooklyn in 1909. By the 1920s, when the first boom of American hospital building had subsided, and when the general hospital had come to assume a prominent place in the medical treatment of people of all classes, the Catholic Church in New York had fourteen general hospitals consisting of a little over 2,600 beds. This was the highest number of general hospitals (and beds) among any benevolent or religious group.[4]

At the earliest Catholic institutions founded as general hospitals—those intended for patients of both sexes and all ages suffering from a variety of illnesses—most patients would be described today as suffering from chronic illnesses. "Old age" was a category of admission at St. Vincent's, and residents at St. Francis' included elderly patients (referred to as grandmothers and grandfathers in a contemporary account) who had nowhere else to live. Patients at St. Vincent's stayed for indefinite periods, some remaining from three to fifteen years.[5]

In 1863, the first year statistics are available from St. Vincent's, patients were treated for diarrheas, dysentery, fractures, fevers, hysteria, and gunshot wounds, but 23 percent of the patients treated were noted as consumptive—what would come to be called tuberculosis after the identification of the bacillus in 1882. There were so many of these patients at St. Vincent's in its first decades that in 1859 the *Metropolitan Record* sought to explain the hospital's high death rate by noting how many patients were admitted in the "final, fatal stage."[6] Other admissions that year included twenty-two patients admitted for debility and two for obesity.[7]

Some Catholic hospitals reflected their founders' specific concerns about immigrant women, and several of the earliest were maternity hospitals that provided homes for needy women before and after childbirth and took in abandoned children. Sisters opened three of these in New York and Brooklyn. The first, St. Mary's Female Hospital in Brooklyn, was founded by the Sisters of Charity in 1868 and reincorporated in 1888 as St. Mary's Maternity and Infant's Home. In Manhattan, the Sisters of Charity opened their New York Foundling Home and Hospital in 1869, and the Misericordia sisters organized the New York Mothers Home of the Sisters Misericorde in 1888.[8]

Given the primarily caretaking nature of the nineteenth-century hospital, Catholic hospitals most often described their superiority in terms of their nursing staffs. *Catholic World* noted in 1868 that in Catholic hospitals there were "no hired nurses; life devotion supplies all."[9] Catholics and non-Catholics alike commented on the New York hospital sisters' exceptional dedication to their work. A typical quote from outside the church in 1872 concluded that "not withstanding all the errors of their faith and practice, [the sisters] present a sublime anomaly in the history of the world, and are eminently worthy of imitation."[10]

In a period when there was no formal educational process for nurses, sisters' dedication and lifestyle had practical manifestations on the hospital ward. As Siobhan Nelson points out, Protestant deaconesses and Anglican and Roman Catholic sisterhoods in the nineteenth century all developed and presented a model of nursing usually exclusively attributed to Florence Nightingale.[11] New York's Catholic hospital sisters were disciplined and careful nurses decades before nursing developed into a trained occupation that called for those attributes. Even their unique dress contributed to their success. Before the hospital training movement initiated uniforms for nursing students and graduate nurses, sisters even looked more capable than other nurses. (Charles Rosenberg quotes a young lay nurse in one of the early hospital training programs reflecting on the advantages of a uniform: "I find my cap and uniform . . . a great help in managing the patients—now that I work in this dress they cant [*sic*] tell how little I know but will obey almost unquestioningly.")[12] The sister's religious habits set them apart, identified them, and gave them reassuring authority. Unlike other hospitals where convalescing patients were called on to care for other patients, Catholic hospitals offered patients the prestige of nursing care by sisters.

While claiming superiority based on the quality of their nursing sisters, Catholic hospitals never provided any physical alternatives to orthodox medical practices. Their religious beliefs coexisted easily with what has come to be called nineteenth-century regular medicine. As a result, they were not embroiled in controversies like those that Christian Science practitioners would face in later decades. Sisters prayed for their patients—indeed, they saw their nursing as a form of prayer—but they never publicly framed the care they offered in terms of miracles exclusively. Moreover, the medicine practiced by physicians in Catholic hospitals, and the physical procedures the sisters as nurses followed, was decidedly mainstream.[13]

The doctors first involved with St. Vincent's in the mid-nineteenth century were members of an elite group of New York physicians also affiliated with other hospitals throughout the city. The first president of St. Vincent's medical board and chief surgeon was Dr. Valentine Mott, also chief surgeon at New York Hospital and Bellevue, the city's only other two hospitals in 1849. Supporters were eager to have his name attached to the St. Vincent's. An article describing the hospital in the *Freeman's Journal* noted how Mott's work was "well calculated to inspire confidence in the treatment at this Hospital."[14]

In the following decades, physicians' practice of having affiliations at multiple hospitals continued. Mott's son was an attending surgeon at St. Vincent's, the New York Dispensary, and Jews' Hospital (later Mount Sinai). His son-in-law William Van Buren was surgeon at Bellevue, New York Hospital, and then St. Vincent's. Physicians like Valentine Mott and his family and friends were anxious to extend their hospital affiliations because the hospital offered them clinical experience and the possibility of financially rewarding teaching opportunities. In the nineteenth century, only a select few physicians received hospital training, and it was a coveted credential needed to establish a lucrative private practice. The physician's ultimate goal was the development of a paying private clientele and a hospital affiliation was a means to that end; the hospital's religious orientation made no difference in that respect. A Dominican sister who was the first pharmacist at Mary Immaculate Hospital in Queens, as well as its historian, described the priorities of physicians and the founders of Mary Immaculate in plain terms: "They didn't care who ran the hospital as long as they got a hospital."[15]

Catholic hospitals proponents rarely suggested that Catholic care would produce miracles. Patients at Catholic hospitals were promised superior treatment because it was Catholic care, but this was not couched as divine intervention. Supporters most often explained Catholic superiority in hospital care with references to the motivation behind sisters' involvement in health care, which they claimed had a pervasive and positive influence in the institution. As the *Metropolitan Record* noted in 1859, St. Vincent's was no different from other hospitals unless "we accept the difference which is made by the fact that in this hospital charity is the ruling motive."[16] An 1862 article explained that the hospital was an "asylum . . . where consoling influences of religion may be had in peace and quiet."[17] Another suggested Catholic hospitals were better than others because in a Catholic institution, "corporal and spiritual needs could be attended to."[18]

Hospital annual reports and booster articles in the Catholic press emphasized that Catholic hospitals did not cater exclusively to Catholic patients and that patients were never under any obligation to attend formal religious services. These explanations resulted directly from protests in which the Catholic clergy complained about the prominent role of Protestant clergy in other hospitals, including municipal ones (see chapter 1).[19] Furthermore, supporters often made the point that Catholic hospitals allowed non-Catholic patients visits from their own "spiritual advisors."[20] Articles frequently noted that all patients in a Catholic hospital were welcome to seek the solace of their own religion. Right after St. Vincent's opened in 1849, the *Freeman's Journal* printed a story to emphasize this point. If the Sisters of Charity hospital in Buffalo was an indication of how Catholics conducted their hospitals, the paper concluded that New York's Protestants had little to fear from the Sisters at Saint Vincent's in New York. Protestant clergy at the Buffalo hospital were allowed to visit patients and conduct funeral services. In other words, those patients who entered a Catholic hospital as Protestant could very openly leave the hospital that way—dead or alive.[21]

On a similar note, in 1879, the *Catholic World* pointed out that St. Francis' Hospital allowed Protestant and Jewish clergy access to patients during hospital visiting hours and as "often as [they] call for them." The paper bragged that when non-Catholic patients were close to death, the sisters always recommended that they call for a clergyman of their own faith.[22] In other New York hospitals, by contrast, "the visits of priests and sisters, if not forbidden, are obstructed and discountenanced."[23] Less enthusiastic commentators noted that while non-Catholic clergy could visit patients, only Roman Catholic services were allowed on the premises.[24]

Despite this rhetoric, these hospitals were clearly Catholic institutions for Catholic patients.[25] Catholicism was physically apparent to patients through more than the presence of nuns. Sisters decorated their hospitals as they ornamented their convents, with plenty of religious pictures and statues. An article in the *Freeman's Journal* in 1858 described a Catholic hospital in Detroit as a virtual Roman Catholic picture gallery. "The first that will strike you is the pleasant face of the Bishop: or it may be the sad expression of the Holy Mother . . . St. Vincent will also claim a share of your observation, as his benevolent face seems to call you."[26] Photographs of St. John's Hospital in the early twentieth century show a children's ward with a statue of the Blessed Mother keeping watch over cribs and rocking chairs. Sisters wanted private rooms to look comfortably domestic and furnished them in keeping with their model of a proper Catholic home; a crucifix hanging on the wall would be visible throughout the room in the mirror hanging over the dresser. As a home might be decorated with family portraits, the walls of their hospital's reception rooms were adorned with portraits of the hierarchy or favorite saints.[27]

Any thoughts on whether all this sectarian décor was agreeable to patients were moot. Charity patients who were sent to a hospital by an agency or organization had little choice about where they would be admitted. A husband and wife both received assistance from a private charity in New York in 1899: the husband who went to Bellevue had been baptized a Catholic, the wife who was at St. Francis' was Protestant. Religious orientation of patient or institution seems to have been inconsequential to the decision, although the truth, according to a social worker from yet another charity institution that was following their case, "was that they have no church connection."[28] That may or may not have been so. What was evident was that their poverty limited their control over their lives.

Economic circumstances or lack of choice with regard to hospitals were not necessarily typical of all other Catholic hospital patients, however. St. Francis' was initially a charity hospital where all the patients were destitute, but at other Catholic general hospitals patients often paid something toward their care. Records indicate that most patients were working-class people—mostly, but not exclusively, unskilled laborers. The majority of patients treated at St. John's Hospital in its early days, for example, noted some form of employment. The largest percentage (36%) of adult patients between 1891 and 1893 was classified simply as "laborers" while fewer than 10 percent were skilled workers. A list of patient occupations at St. Francis' Hospital in 1905 is similar; it includes laborers and domestics but also some shoemakers and seamstresses. At St. Vincent's in Manhattan, most patients paid something toward the cost of their care, suggesting a patient population not only willing but able to contribute something, even a minimal amount, to the cost of their care.[29]

Patients at a Catholic hospital often shared ethnic or national identity as well. These hospitals originally had been founded for Catholics, albeit nominal ones, and specific ethnic groups within the Catholic population. Annual reports often listed patient nativity but not religion and indicated that patients at any given hospital overwhelmingly shared national identities. The *Metropolitan Record* pointed out in 1868 that while St. Francis' Hospital accepted patients of all nationalities, it had a "more than usual average of Germans."[30] In 1892, the New York State Board of Charities noted that Columbus Hospital had always offered free care to all the "worthy sick poor" although "more especially . . . Italians."[31]

Catholic hospitals maintained their ethnic affiliations throughout the period even as their foreign-born populations dwindled. St. Vincent's patient rosters between 1890 and 1900 indicate a strong, yet slowly dwindling, immigrant and predominantly Irish population. The percentage of St. Vincent's immigrant patient population is very consistent from 1888 until 1894. For most years with available information, the foreign-born population accounted for an average of 64 percent of the total population, with Irish-born patients always comprising the greatest number of the foreign-born. But by the turn of the century, St. Vincent's population started becoming native-born: in 1899 there was a notable rise to a native-born majority, which was repeated the next year. Irish-born patients continued to account for 25 percent of the patients treated at St. Vincent's, the largest number of any immigrant group. St. Francis' Hospital maintained similarly strong ethnic affiliations. In 1905, the number of German-born patients was higher than the number of native-born (although not by much) and was more than any other immigrant group.[32]

Catholic hospitals were located within immigrant neighborhoods and did not usually begin life with new construction. Although nineteenth-century medical experts noted that "a well planned hospital would be carefully located in a healthful site," hospital founders usually had little choice about location."[33] In most cases, Catholic hospitals were initially renovated from existing and available private structures. While the Sisters of Charity had hoped to be able to build a new structure on the East River for their first hospital, St. Vincent's opened in a rented house on 13th Street in Manhattan—a much less expensive arrangement. Similarly, St. Francis' Hospital was originally housed in two connecting buildings with the wall between them torn down.[34] While city health officials noted the hospital was in a particularly unhealthy part of the city, it was convenient for the German immigrants it cared for.[35]

The size of Catholic hospitals also made them less overwhelming and less intimidating than the massive city-run institutions. Even as Catholic hospitals increased in size, they remained much smaller than the hospitals operated by the public authorities. In 1892, New York City's Bellevue Hospital had 800 beds; City Hospital on Blackwell's Island had 1,000; King's County Hospital in Brooklyn could maintain 400 patients at a time. In contrast, St. Vincent's capacity was 170, St. Francis' was 240, and St. Catherine's was 180.[36]

Although Catholic institutions offered less intimidating surroundings than most other hospitals, sisters, like other hospital administrators, sought to impose a particular mode of behavior on patients. Charles Rosenberg has described nineteenth-century hospitals as a "battleground of values," where administrators continually fought with working-class immigrant patients who objected to attempts to monitor and control their behavior during their hospital stay. Given the sisters' religious and cultural affinity and their status within the Catholic world, nothing quite that confrontational took place in Catholic hospitals. However, sisters had rules about behavior too.[37] While the hospitals Rosenberg describes might have been trying to make their patients middle class, sisters wanted to show them what a good Catholic life was all about. In many ways they were the same thing. The Catholic life of hospital sisters was disciplined, and sisters sought to establish order and decorum on patients by limiting the frequency of visitors and establishing rules. Patients were advised that there was to be no sitting on the side of beds, talking aloud, or smoking. Ward patients were not allowed to leave the ward and enter any other hospital room or facility, with the notable exception of the chapel, without permission. The patient regulations at Seton Hospital, a tuberculosis hospital run by the Sisters of Charity where the patients were almost entirely charity cases, illustrate how the sisters attempted to maintain what they considered propriety, and demonstrate that class distinctions among patients and staff were not absent from Catholic institutions. In addition to stringent regulations established to avoid tubercular contamination, the patients were also reminded that they were not allowed to gamble.[38]

The growth in the number of Catholic hospitals in New York and Brooklyn was part of a national phenomenon. Hospitals, especially in urban areas, increased dramatically after the Civil War. These new hospitals were increasingly organized as general care hospitals, and some hospitals founded for other purposes were reorganized as general hospitals. Turn-of-the-century general hospitals differed from earlier ones: they were more likely to treat patients suffering from what are better described as acute conditions, which could and would be treated immediately with chance of recovery, rather than chronic ones requiring long-term care. They also described themselves less frequently as charitable institutions.[39]

In some ways, the Catholic hospital story in New York City seems to conform to this pattern with regard to both growth and development. Unlike the general hospitals established earlier in the nineteenth century, St. John's Hospital, founded in 1891 by the Sisters of St. Joseph, was from the very beginning an institution that cared for patients needing immediate and short-term care. A sample of the patient population between 1891 and 1894 indicates that the most common admissions at St. John's were assorted fractures (21.5%), followed by contusions (11.3%), lacerations (10.25%), amputations (6.8%), and burns, scalp, and other wounds (5.6% each). St. John's location in industrial Long Island City was a strong factor influencing its medical orientation. The categories of patients admitted to St. John's, and the fact that most were male (86.3%), suggests that the majority of admissions at St. John's were work-related injuries. Many of the patients were railroad workers.[40]

At the same time, St. Vincent's, which since its 1849 inception had seen mostly long-term cases, was undergoing a shift away from chronic care. The number of patients diagnosed with chronic tuberculosis at St. Vincent's declined significantly between 1880 and 1900.[41] At the same time, both the number and kinds of operations performed at St. Vincent's increased. The earliest figures available on operations are for 1881 and show few performed that year: 3 percent of the patients treated were operated on. The operations included incisions, amputations, bone setting, bullet extractions and there was one excision of the breast.[42] By the end of the decade, the number of operations performed increased, reflecting the post-Lister asepsis technique that significantly lowered the infection rate. In 1888, surgeons at St. Vincent's continued to set bones and amputated fingers but also removed a variety of tumors. In the 1890s, gynecological cases and operations increased and received separate categorization in annual statistics. By 1906, operations were more likely to be performed on female patients than male: an almost two-to-one ratio.[43]

The language used by hospital founders to describe their work reflected these changes. St. Vincent's was organized "for the care and treatment of the indigent sick," but St. John's Hospital in industrial Long Island City was established in 1891 to "care for and maintain the sick and injured" and made no reference to the patients' economic status. Similarly, Mary Immaculate Hospital, which opened in Jamaica, Queens, in 1904, was organized simply for "medical and surgical treatment."[44]

Mission statements and even patient statistics do not tell the entire story of what was happening to Catholic health care in New York at the end of the nineteenth century. In some ways these Catholic hospitals do not conform to the pattern of development going on in other hospitals because Catholic hospitals were not monolithic institutions. Rather, they were pieces in a mosaic of institutions and services provided by Catholic sisterhoods; to consider their history as independent institutions is misleading. If we recognize that Catholic hospitals were one part of a larger landscape of health care and social welfare services, which is how the sisters intended it to be and worked to make it, then the history of Catholic hospitals is not entirely identical to that of other hospitals.

As their general hospitals shifted their focus from chronic to acute care, sisters opened new institutions that expanded and continued their original hospital work—their involvement with mothers and abandoned children, and consumptive and aged patients. In Manhattan, the Sisters of Charity expanded the New York Foundling (alternately referred to as the Foundling Asylum, Foundling Home, and Foundling Hospital) and opened new institutions, Seton and St. Lawrence Hospitals, to care for consumptives, now referred to as patients with tuberculosis. In opening these institutions, sisters were once again following mainstream medical practices, participating in the sanatorium movement that separated tuberculosis patients from others.[45]

The Sisters of St. Dominic opened Our Lady of Consolation Residence as a convalescent home for patients from St. Catherine's. In 1892, Consolation was reorganized as a home for the elderly, presumably because they made up the majority of the institution's population. The

Sisters of Charity also maintained a home for the elderly, St. Joseph's Home for the Aged, which they organized in 1868. When the Franciscan Sisters moved their hospital from lower Manhattan to East 142nd Street in 1905, they maintained a home for the aged and chronically ill at the old location.[46]

Despite the other changes at St. Vincent's at the end of the century, one category of admission requiring chronic care remained at a high and stable percentage. Patients categorized as alcoholic continued to outnumber any other classification admitted to St. Vincent's. In 1881, 192 of the hospital's 1,962 patients were noted as treated for alcoholism. High numbers continued in the following decade: 14 percent in 1888, 12 percent in 1894, 13 percent in 1900 and 14 percent in 1906. Annual reports distinguished between chronic and acute cases of alcohol "poisoning." The hospital continued to be a custodial institution at least for patients classified as alcoholics.[47]

Sisters, and their hospitals, are an important part of the story of the development of the modern hospital in New York. Early interpretations explaining the rise of the new general hospital stressed scientific advancements and improvements in medical techniques; more recent historiography has noted the equal significance of other factors, including public perceptions about hospital care. It is arguably in that area—what people came to think about hospitals—where sisters made their greatest contribution.[48]

New York's first Catholic hospital opened when hospitals were characterized in primarily negative terms. The *Metropolitan Record* remarked in 1859 that St. Vincent's was not "understood by those whose benefit it was most intended," reflecting contemporary fear and dissatisfaction with hospital medical care.[49] St. Vincent's and later Catholic hospitals countered the image of hospitals as frightening, unfamiliar, and uncomfortable, not to mention unhealthy and hostile, places. The "active sympathy and unruffled composure" of the hospital sisters provided immigrant Catholics with an alternative model of hospital care.[50]

## CHAPTER FOUR

# "Building in New York Is Very Expensive"
## *Hospital Finances*

Sarsaparilla, ducks, lambs, turkeys, daily groceries, and ten dozen spools of cotton are just a few of the gifts from patrons of one Catholic hospital in 1888.[1] Contributions like these were typical and reflect the kind of relationships on which hospital sisters relied to sustain their institutions. Sisters were enormously successful in creating a network of supporters among New York's immigrant Catholics. Any image of sisters as passive and sheltered women crumbles when we see the resources they cultivated. Sisters had a keen understanding of their likely supporters.

The simple beginnings of their hospitals—a few sisters moving into a small building—obscure the critical decision sisters made when they chose to open a hospital. While both the Archdiocese of New York and the Diocese of Brooklyn encouraged sisters to open hospitals, neither promised or assumed any permanent financial responsibility for them. Any costs sisters took on—a lease, a mortgage, or daily expenses—belonged, with few exceptions, to them.

Realizing that the ultimate financial responsibility for a hospital was theirs, sisters did not make reckless decisions and simply go wherever they were asked. Because of financial concerns, the Sisters of Charity in 1843 declined Bishop John Hughes's first suggestion that they open a hospital. Recognizing that sisters would never take on a new hospital without some solid backing, the Brooklyn bishop told residents of Far Rockaway in 1904 that if they wanted a Catholic hospital built there, they would have to come up with some money first; he would not ask sisters to consider the idea otherwise.[2]

In the nineteenth century and for much of the twentieth, Catholic hospitals had a unique relationship with the church hierarchy. The hospitals were not owned by either the Archdiocese of New York or the Diocese of Brooklyn, the Roman Catholic administrative units in which they were located. In a vague and peculiarly Catholic way, hospitals were always under the supervision of the bishop who headed the diocese, but what that meant in practical terms was never spelled out.

The financial organization of the Roman Catholic Church in the United States is a complicated and thorny affair; it differs among dioceses and is dictated by civil statute as well as canon law. Some dioceses, Chicago for example, were initially organized under an unusual nineteenth-century corporate structure called "corporation sole," in which the bishop personally became a legal corporation who owned all diocesan property. A more widespread method of church organization and financial management, and the one used in both the Archdiocese of New York and the Diocese of Brooklyn, was the "corporate aggregate" system. Under that method, developed in New York in the 1840s by Bishop Hughes, all Catholic organizations and institutions in the diocese, including hospitals, were organized as separate corporations. The hierarchy frequently controlled the individual corporations as holder of the majority of seats on the corporate boards.[3]

In both the corporation sole and the corporation aggregate systems, the corporate structure of the hospitals and related provisions of canon law prohibited the hierarchy from redirecting hospital funds or properties to other purposes. Even in dioceses where the bishop had the very powerful privilege of corporation sole, hospitals remained outside his purview. As Edward Kantowitz explains in his history of the Archdiocese of Chicago, hospitals eluded episcopal control. As far as hospitals went, usually all the bishop could do was "inspect them, and raise hell in Rome if he didn't like what he saw."[4] The same was at least literally true in New York, although connections between some hospital communities and the hierarchy were close. At the same time, the hierarchy was not responsible for the financial maintenance of Catholic hospitals, and this corporate arrangement created a formidable distance between the hierarchy and the monetary needs of the hospitals.[5]

Hospital boards included members of the hierarchy and other clergy, sisters, and laymen, but the hierarchy did not have majority control of the hospitals' boards. While the bishop and other clergy were often trustees, sisters had numerical superiority in many hospitals. St. Vincent's Hospital in Manhattan was originally organized under the Corporation of the Sisters of Charity of New York. That corporation was extended in April 1857 to "purchase land and buildings, and to erect buildings for the purposes of a Hospital in the City of New York." As directors of the Corporation of the Sisters of Charity of New York, the archdiocesan vicar general and the bishop were also directors at St. Vincent's, but they were outnumbered on the board of managers by sisters. Similarly, nine Sisters of Charity signed the articles of incorporation for Seton Hospital, a tuberculosis hospital that opened in 1894. At St. Elizabeth's Hospital, Monsignor Joseph Mooney, vicar general of the New York Archdiocese was president of the boards and all the officers were Franciscan sisters. At St. Joseph's in Far Rockaway, the directors were Rt. Rev. Joseph McNamee of the Brooklyn diocese and Mary Ann Crummey, Mary Ann Mahoney, Marcella Gill, Mary Ennis, Sarah Boylan, and Mary Pollard—all Sisters of St. Joseph.[6]

The arrangement between sisters and clergy in hospitals was very different from the Catholic parochial schools, which sisters also staffed and managed. Unlike the schools, most Catholic hospitals did not have specific parish affiliations. With two exceptions, St. Francis' Hospital in Manhattan and St. Catherine's in Brooklyn, Catholic hospitals were not clearly organized for the benefit of one parish or connected either financially or physically to a parish. (St. Catherine's connection to Most Holy Trinity parish was reflected in its corporate structure: the pastor of Holy Trinity was the vice-president of the board of managers.)[7]

In contrast, most Catholic elementary schools were organized within individual parishes to serve its parishioners. Sisters came to staff and administer parish schools (and sometimes parish orphanages), after a pastor asked them to. When he did, financial details of the responsibilities of all parties were clearly specified. When Father Lewis, the pastor at St. Mary's in Manhattan, asked the Sisters of Charity at Mount St. Vincent's to send a group of sisters to his parish in 1867, he specified to the community's superior general what the financial arrangements were to be. In order "to prevent misunderstanding," his memo made it clear that he alone had "the administration of . . . financial affairs. The sisters keep the accounts, provide whatever is necessary and at the end of every month, Father Lewis gives them the amount to their bills, as per the books."[8]

Financial independence could be a less than desirable situation if money was scarce, and in the early days of Catholic hospital development, it usually was. The individual communities of sisters were all separate corporations, and none were well endowed with either land or accumulated capital. There was also very little incoming money when a group first came to New York. Unlike many earlier European sisterhoods, New York communities did not have the benefit of large dowries, the money and property women brought with them into the convent. Their only source of steady income before they opened hospitals and other institutions, was tuition money, either from parish schools or from the more expensive female academies almost every community ran. Parish schools could not always be counted on for income; as the Dominican sisters noted in 1858, "more than half of the children are unable to pay tuition, our only source of income, due to the unemployment of their parents."[9] In the academies, where sisters offered a genteel Catholic education to young women with economic means, tuitions were higher and hopefully more reliable, but they could never completely cover what quickly became increasing hospital costs.[10]

The single most important factor contributing to these rising costs was that almost immediately after opening, sisters saw a need to expand their hospitals, and as a Sister of Charity at St. Vincent's wryly noted in 1853, "Building . . . in New York is very expensive."[11] Over the course of fifty years, St. Vincent's population increased from 299 patients treated annually to more than six thousand. The Sisters of Charity spent $21,109 to maintain 800 patients in 1863; in 1910 it cost $228,776 to treat six thousand patients.[12]

Sisters were not cavalier about the scope of their efforts. Concerns about how far a community could and should extend itself sometimes limited the size of a venture. The Sisters of Charity originally planned to erect a new building to house St. Vincent's but instead rented an existing building, presumably at a much lower coast. When the Sisters of St. Joseph bought land for St. John's Hospital in Long Island City in 1891, they purchased several partially completed buildings on the premises, but without money to renovate them, they could not use all the buildings immediately.[13]

As hospitals became larger, they would assume a dominant physical place in a neighborhood. St. Vincent's, just a little house when it opened, grew to be a significant presence in Greenwich Village by the turn of the century. Just three years after the hospital opened, the Sisters of Charity rented a second building on the same street. Soon afterward, in 1856, they moved into larger quarters across town on West 11th Street and rented a building that had earlier housed St. Joseph's Half Orphan Asylum. It was at this address that the hospital would become a neighborhood fixture. In 1868, they bought the building and started accumulating other properties nearby, purchasing houses on West 12th Street in 1863 and 1874. In 1892, they bought a nearby synagogue. (Expansion often brought internal improvements which could also be costly. The first building had no gas light or internal plumbing; by the 1870s, the hospital had steam heat and hot and cold baths.)[14]

The Franciscan sisters' work at St. Francis' Hospital progressed similarly, with the purchase of a nearby house just one year after opening. In 1869, the hospital was enlarged further through the addition of three other buildings; another was added in 1871. By October 1875 when construction and renovation was completed on all existing buildings, St. Francis' could house 150 patients; by 1884, with yet more additions, its capacity was 280.[15]

St. Vincent's expenses in 1863 demonstrate the high cost of enlarging the hospital (Table 4.1). Little of nineteenth-century medical cost was actually spent on medicine: food costs totaled 40 percent of monies spent that year. Other maintenance costs, including rent, repairs, fuel and lighting, amounted to less than 15 percent of the total expenses. Hospital expansion accounted for the most money spent. The largest expense—almost one-third of total expenditures that year—was the $7,500 the sisters put toward the purchase of a new building.[16]

Sisters worked hard to make the most out of their resources—community records include carefully noted estimates on building and renovation costs—and they made sure to emphasize this frugality in efforts to bolster financial contributions from both private and public sources. At the same time, they were careful to note they did not skimp when it came to patient care. The Sisters of St. Joseph's at St. John's Hospital were typically humble but deliberate in the wording of their annual report in 1892 when they noted that "The furnishing of the hospital is simple but durable, in keeping with the uses intended . . . The groceries, meats, liquors, drugs etc., are purchased from first-class firms only, and of best quality, it being deemed wiser to pay for a good article rather than be misled by what might prove false economy."[17]

Sisters' assessment of financial realities was helpful to them as they searched for ways to come up with the money they needed to open and run hospitals. Unlike some other private hospitals in New York, none had the initial support of a wealthy benefactor who financed the initial venture.[18] The Sisters of Charity supplied all the starting funds for St. Vincent's in 1849, but that was not always the pattern for their community or others. The initial sources of hospital funding varied among Catholic hospitals. In a few cases, sisters began hospital work with the help of local clergy. More frequently, however, sisters got started in hospital work with significant help from lay Catholics. Catholic men and women contributed small and large amounts of money, services, furnishings, food, and their own time to begin hospitals and, later, to keep them afloat. How sisters garnered this support and why lay Catholics helped them is a story of effective leadership in both health care and community development. Sisters proved themselves remark-

*Table 4.1* Contributions from the Church of the Most Holy Redeemer to St. Francis' Hospital, 1865–1868

| *Source* | *Amount* |
|---|---|
| Collections in the Church of the Most Holy Redeemer | $1,037.00 |
| Collections by the Fathers in private houses | 3,006.52 |
| Rev. F. Braidstutter extra collection | 4,000.00 |
| Monthly collections by the Women of the Society of the Holy Family | 2,487.58 |
| Two excursions under the auspices of the Independent Rifle Company | 2,276.00 |
| Individual donation | 2,500.00 |
| Individual donation | 1,000.00 |
| Appropriation from the state | 18,000.00 |
| Loan from the Societies and from the Church of the Most Holy Redeemer | 6,348.84 |
| Loan from the Societies of St. Alphonsus Church | 3,608.00 |

*Source:* "Claims of the Fathers and the Congregation of the Church of the Most Holy Redeemer, New York, to the St. Francis' Hospital 5th Street New York," Archives of the Archdiocese of New York.

ably skillful persuading would-be contributors that theirs was a cause worth supporting.

A great deal of the initial fundraising work for Catholic hospitals was done by Catholic lay women at fairs organized to support individual hospitals. As Colleen McDannell has illustrated, ladies' fairs were a frequent fundraising vehicle among Catholic New York lay women in the nineteenth century, and hospital fairs were similar to the parish fairs she describes. More than just fundraisers, they were popular as social events as well. There was music and food and all sorts of items to buy, even for those with little money to spend. Raffles promised even more—a piano, a diamond brooch, or a silver tea set.[19] With typical flourish and bravado, the *Freeman's Journal* described the first of these hospital fairs, held in 1856 for St. Vincent's, as "an event in the history of New York."[20] The fair netted approximately $35,000 and provided a nice nest egg for a few years. In 1863, fair proceeds were still being applied to outstanding bills: the $3,600 still remaining was of significant help, covering 17 percent of the approximately $21,000 in hospital expenses that year.[21]

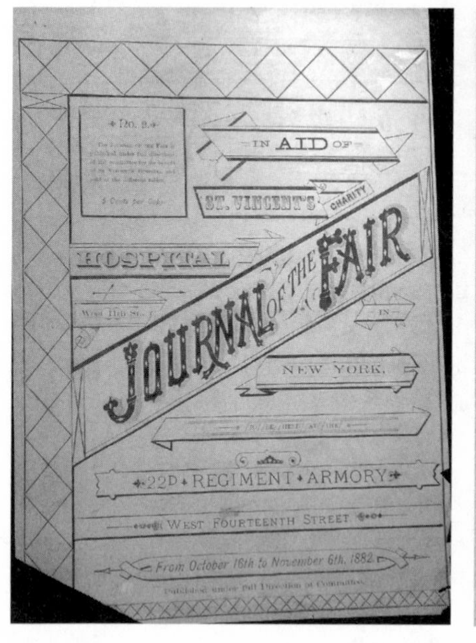


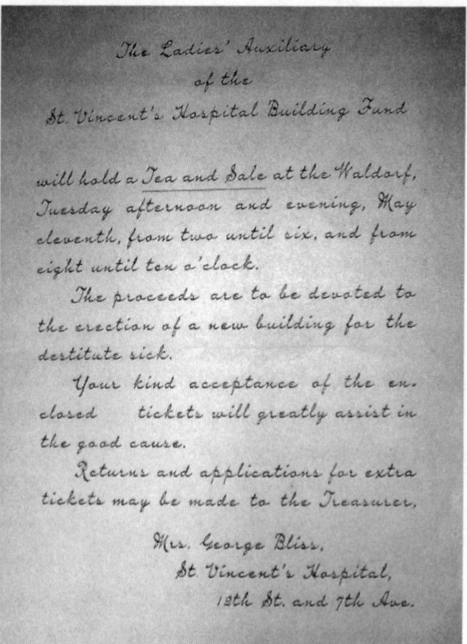

The Ladies' Auxiliary
of the
St. Vincent's Hospital Building Fund

will hold a Tea and Sale at the Waldorf, Tuesday afternoon and evening, May eleventh, from two until six, and from eight until ten o'clock.

The proceeds are to be devoted to the erection of a new building for the destitute sick.

Your kind acceptance of the enclosed tickets will greatly assist in the good cause.

Returns and applications for extra tickets may be made to the Treasurer,

Mrs. George Bliss,
St. Vincent's Hospital,
12th St. and 7th Ave.

*Fig. 7.* Although only fifteen years apart, the ca. 1897 invitation to a tea at the Waldorf organized by a newly formed Ladies Auxiliary reflects a shift from the fairs (*Journals of the Fair,* 1882) of the previous decades to more genteel women's fund-raising events in the twentieth century. Sisters of Charity of New York

The success of these fundraisers was largely due to women's efforts: women took charge of planning and running most of the events. Their fair responsibilities involved women in tasks and responsibilities outside their homes and family, but this work did not threaten gender boundaries. Charity work fell well within accepted female spheres of influence and involvement. More noteworthy is that Catholic women, religious and lay, worked together to fulfill the sisters' mission to health care.

Unlike church fairs, hospital fairs transcended the geographic boundaries of parishes. Fairs were joint efforts by several parishes brought together through and by a community of sisters. Women from several Manhattan parishes worked on the 1856 fair for St. Vincent's. The booths, decorated with needlework and household articles and trinkets, were designated by parish name, and there was probably no small amount of friendly competition. Profits were noted by parish name and ranged from the $750 raised at St. Columba's booth to nearly $2,500 from St. Patrick's. (St. Patrick's had not yet moved to Fifth Avenue but was still the seat of the diocese.) Ethnicity was the critical factor in terms of involvement. All the parishes involved with St. Vincent's were Manhattan's Irish ones. German-Catholic women supported their own hospitals with similar efforts. Fairs for St. Catherine's Hospital raised almost $23,000 between 1869 and 1873.[22]

Personal ties between sisters who ran hospitals and the Catholic community also helped determine who would support a particular hospital once it opened. Some hospitals had special patrons who made sizeable and frequent contributions. In these cases too, a close relationship between the donor and the hospital sisters often precipitated the gift. Eugene Kelly, related by marriage to both the hospital's founder Ellen Hughes and her brother Archbishop John Hughes, was a close "friend and advisor" to St. Vincent's. He endowed several beds for the use of charity patients and also made cash donations, including $10,000 in 1893, the year before he died.[23] Like Kelly, other large contributors had family connections to the sisters. The father of one of the Dominican sisters at St. Catherine's made one of the few hefty ($5,000) individual donations to their hospital. When St. John's in Long Island City opened, the father of a Sister of St. Joseph donated a building for the sisters to use as a convent.[24]

Physicians were also financial supporters of Catholic hospitals, but for different reasons. As in other nineteenth-century hospitals, physicians in Catholic hospitals did not receive any payment for their services. This was not entirely a charitable contribution on their part. As historian Morris Vogel explains, this system worked to the physician's ultimate financial benefit. While nineteenth-century physicians earned their living from payments received from private (as opposed to hospital) patients, hospital work and experience contributed to the reputation that brought paying patients to their offices. "Without this gratuitous service," Vogel notes, "it was difficult for a young doctor to begin a practice in a city where paying patients had a wide choice among practitioners and would choose experience."[25]

In addition to their gratuitous services, physicians made other contributions. In 1904, one doctor helped the Dominican sisters with the rent on the building that first housed Mary Immaculate Hospital. John A. Harrigan, a physician and president of the board of trustees at St. Mary's Hospital, made several substantial gifts to that hospital over the course of his thirty years there. In addition to cash gifts of $10,000 and

$6,000, he contributed more personal and more visible gifts, supplying patients and staff with special holiday dinners, for example.[26]

Priests, parish organizations and individual lay people were all early contributors and fundraisers at St. Francis' Hospital (Table 4.2). Between 1865 and 1868, two churches, Most Holy Redeemer and St. Alphonsus, loaned the Franciscan sisters almost $10,000. Male and female parish organizations, like the Holy Name Society, contributed separate amounts totaling almost $5,000. Two parishioners made significant individual donations: one $1,000 and the other $2,500. Another $8,000 was raised by special collections held in church and through clerical solicitations. By 1868, when the hospital's total yearly expenses were approximately $36,000, Most Holy Redeemer had raised over $40,000 for St. Francis'.[27]

St. Francis' was unusual in its specific parish support and also in its specifically German affiliations. Like the parish of the Most Holy Redeemer, the hospital was organized to meet what were perceived to be the unique needs—notably language—of German Catholics in New York. St. Catherine's Hospital in Brooklyn was similarly connected to one German parish: Most Holy Trinity. The majority of hospitals, however, were organized by sisters with connections to a number of different parishes—most of which were Irish ones.

Many hospitals tried to set up systems of guaranteed support through a program of endowed beds. Individuals or organizations could pay in advance for a bed, which would be filled by a patient of their choice when necessary. Potential donors were asked to contribute to a collective fund where members each paid a part of the cost. In 1863, St. Vincent's invited patrons to "form clubs of twelve persons each—each member subscribing ten dollars; and each club thus formed securing one free bed for a year at the reduced price of $120.00 per annum," suggesting this procedure to "Benevolent Societies who . . . are obliged to take care of their members during their illness."[28] As one of the hospital's annual reports explained, not only did membership entitle subscribers to the use of a hospital bed, but the members of the "Beneficial Association of St. Vincent's Hospital" also received the added benefit of "the prayers of the sisters and the sick poor."[29] But subscription fees did not increase much over time and were never a major source of hospital income. In 1867, St. Vincent's raised only fifty dollars through subscription; in 1895, annual subscriber fees totaled $1,044.75.[30] When St. John's Hospital in Long Island City opened in 1890, the local St. Vincent de Paul Society, Exempt Firemen's Association, and Ancient Order of Foresters all bought hospital subscriptions at fifty dollars each, but none established a major precedent.[31]

*Table 4.2* Expenses at St. Vincent's Hospital, 1862–1863

| *Expense* | *Amount* |
|---|---|
| Rent | $1,200.00 |
| Beef, mutton, etc. | 2,768.85 |
| Flour, farina, crackers | 1,219.46 |
| Sugar, tea, coffee | 1,328.22 |
| Vegetables, spices | 832.49 |
| Fish, eggs, butter | 995.18 |
| Milk | 1,044.94 |
| Fuel | 1,228.04 |
| Medicines | 651.78 |
| Liquors, wines, porter | 743.56 |
| Dry goods and bedding | 861.51 |
| Hardware | 281.08 |
| Repairs | 810.38 |
| Wages | 1,041.00 |
| Gas light and candles | 372.90 |
| Croton water tax | 144.00 |
| Insurance | 41.00 |
| Stationery and printing | 118.63 |
| Sundries | 112.81 |
| Purchase of house on 12th Street | 7,500.00 |
| Total | 24,109.12 |

*Source:* St. Vincent's Hospital, *Annual Report 1863*.

At the end of the century, St. Vincent's did better with perpetually endowed beds, which were much more expensive. Some were endowed for the life of the donor, some provided free care for just one year, and others were noted as on account. St. Vincent's first perpetually endowed bed had been established in 1893, but there were seventeen by 1900 and forty-one by 1907.[32]

The increase in individually endowed beds at the end of the century reflects the ability of at least some Catholic New Yorkers to make significant contributions. Catholics expressed their support in other ways, too, frequently by more personal means and often in much smaller amounts. Donor lists at St. Vincent's note contributions ranging from five to five thousand dollars. Other contributions, like Dr. Harrigan's at St. Mary's, were in goods rather than cash. They included small amounts of food and household supplies, as well as larger gifts of horses, wagons, and personal services. These donors were also listed in annual reports, which often specified gifts. In 1863 the sisters at St. Vincent's in Manhattan noted, "Besides voluntary contributions in money, acknowledgements are also due to several friends for donations in stores, and various articles; for all which the sisters return their sincere thanks, and will ever gratefully remember the donors."[33] St. Mary's annual report in 1886 listed gifts of carpets, linens, coffee, and bananas, remarking in particular that "in our capacious kitchen the excellent range and boiler was donated by the late Mr. William Beard." In addition to recognizing patrons, donor lists were also intended to encourage others to contribute. The sisters hoped that some, like Mr. Beard, might make provisions for the hospitals after their deaths, and bequest forms were often included in the reports.[34]

Donors' choices of gifts suggest the place a hospital and its sisters inhabited in the lives of many Catholic New Yorkers. Sometimes gifts were seasonal and celebratory. A benefactor at St. Mary's Hospital in 1886 provided the hospital staff and patients with a traditional Thanksgiving dinner; another at St. Vincent's (future Sister of Charity Euphemia Van Rensselaer) treated patients to ice cream. In 1892, the sisters at St. Vincent's thanked a woman who "made Christmas week especially bright and pleasant for our patients by providing for them a Musical Entertainment . . . followed by a feast of cake and ice-cream for all at her expense."[35]

Large contributors often donated gifts that made a statement about who they were. The largest gifts often reflected a donor's specific interest or occupation or gender. In donations for St. Vincent's new wing in 1891, women took charge of furnishing the wards, a man contributed what was needed for the parlor, and a physician and his wife outfitted the operating room. Each gift reflected appropriate nineteenth-century decorum and social place. The wards were the charity beds and, as such, the appropriate focus for benevolent women. The parlor was the most public of the hospital's rooms, where business was conducted and where the hospital put on a public face—in other words, a suitable location for a man's contribution. The operating room was more than an interest of a generous physician; it was also the place in the nineteenth-century hospital where a doctor's authority was strongest.[36]

Unlike large and small gifts acknowledged in annual reports, many contributions were anonymous. Their neighbors might not know who these donors were, but the sisters who took their contributions certainly would. These contributions, monies that went literally from the hand of the donor to that of a sister, suggest more than anything else the place and power of sisters among the city's immigrant Catholics. Sisters were very successful in their ability to tap into all kinds of Catholic pockets, even ones that were not very deep.

The Franciscan sisters, the Dominicans, and the Sisters of St. Joseph all sent sisters out on missions to solicit contributions. Sisters visited and asked for contributions at places—like police stations and the docks—where they knew they would find Catholics. While these spontaneous contributions the sisters garnered were, for the most part, small, they added up. Between 1873 and 1875, the Dominican sisters who ran St. Catherine's Hospital collected $26,000 through solicitations. By 1892, their success began to alarm Rev. Michael May, the pastor at Most Holy Trinity, whose parish had helped found St. Catherine's. May worried that the Dominicans, who had other charity work in addition to St. Catherine's, might not have been using the money exclusively for the hospital, and they probably were not. Unlike May, the Dominicans viewed all their work as related and did not concern themselves with territorial issues. Also, these donations went from the public directly to the sisters, not to Holy Trinity.[37]

Lay Catholic men supported the sisters through their affiliations with Catholic organizations too. Both the St. Vincent de Paul Society and the Ancient Order of Hibernians furnished wards at St. Mary's.[38] Overall, however, in the nineteenth century, lay men tended to be more influential as advisers, whether as members of the hospital corporation or, as in the case of St. Vincent's, as members of a special advisory board. In many cases these individuals were large donors as well. Eugene Kelly, who gave generously to St. Vincent's, was a member of hospital's advisory board until his death. Others also made frequent contributions and could be relied on for emergency funds. In 1893, for example, board member William Iselin covered the expenses "to care for and to bury a child whose back was broken."[39]

These advisers were also at the forefront of efforts to secure public funding for Catholic hospitals. Advisers were solicited for their political connections as well as their ability to make donations.[40] Many privately organized hospitals in New York, including many Catholic hospitals, received public funding in the nineteenth century. Unlike Catholic schools, which were for the most part denied public financial assistance after 1842, Catholic hospitals received funds from both state and local purses. After the Civil War, when the number of private charity institutions receiving public aid increased, the issue of public funding of Catholic institutions, including hospitals, became a frequent focus of political debate.

New York State's financial involvement with Catholic hospitals began early in the nineteenth century when the state legislature granted funds to private hospitals as reimbursement for the care of indigent patients. The first Catholic hospitals in New York State received appropriations under this system. In 1849, for example, the legislature appropriated nine thousand dollars for the Sisters of Charity's hospital in Buffalo. However, St. Vincent's, the Sisters of Charity's hospital in New York City, which opened at the end of 1849, did not receive any state funding until after the Civil War. Unlike the situation in Buffalo, where the sisters ran the only hospital in the city, New York had several charity hospitals maintained by public authorities.[41]

During the war the amount of public funds appropriated to private hospitals, including sectarian ones, increased as the state paid those institutions for their care and treatment of wounded and sick New York soldiers.[42] After the war, as the number of private hospitals grew across the state, state grants to private charities, including hospitals, continued to increase in size and proportion.[43]

The procedure for state funding was unsystematic and most often based on political connections. Catholic hospitals were eligible as charities, and those in New York City and Brooklyn began to receive public monies when the infamous Boss Tweed of Tammany Hall gained power in state politics. Tammany politicians used their influence to obtain state grants for a variety of favorite charity organizations and institutions, hospitals among them. The first Catholic hospital in New York or Brooklyn to receive a legislative appropriation was St. Francis' Hospital in Manhattan, which received $3,735.52 in 1868. Between 1868 and 1870, all the Catholic hospitals in New York and Brooklyn received legislative grants. They ranged from $713 for St. Mary's Female Hospital in 1870 to $9,950 for St. Francis' in 1869.[44]

For most of the century, New York City aid to private hospitals was no less unsystematic or political than the state procedure. There were two different sources of public funds available to private hospitals. Both the Common Council and the Board of Supervisors could allot funds, but both were ultimately dependent on state authorities since the state legislature had to approve the city's budget. The earliest city appropriation to a Catholic hospital was a thousand-dollar grant to St. Vincent's by the Common Council in 1863.[45]

The question of public aid to private charities became a major public issue at the New York State Constitutional Convention in 1867. Opponents argued that since there were public charity institutions, these private ones duplicated services and any appropriations to them wasted money. There was also mention that public funding of sectarian institutions violated the sacred American principle of the separation of church and state. Some voices were particularly opposed to funding of Catholic institutions—an editorial in the *New York Observer* was entitled "Our State Religion: Is It Roman Catholic?"[46]

Unlike the controversy surrounding state aid to New York City private schools in the 1840s, attempts to reform the state procedure for charity appropriations put the Catholic Church on the defensive. Because the church was already receiving public funds for hospitals, it was not looking to break the monopoly of another private group receiving all public money as had been the case with Catholic schools. As early as 1850, the church had warned that an attempt to discontinue state appropriations to the Sisters of Charity's hospital in Buffalo could have serious consequences for legislators at the next election. New York's Catholic newspaper vowed to print the names of all state legislators who opposed the appropriation for this hospital, the first Catholic hospital in the state, urging their readers to take note for election day.[47]

Catholics moved quickly to prevent changes in the system. An editorial in the *Metropolitan Record* cautioned the paper's Catholic readership that the convention might discontinue further state appropriations to sectarian charities. The editorial dismissed accusations that Catholic institutions received an unfair proportion of funds, noting that "gross misrepresentations on this subject have been made."[48]

The debate continued throughout the convention. One proposal attempted to limit public appropriations to private charities that were not "religious or sectarian in character, and that a majority of its managers are not members of one religious denomination."[49] Such a qualification would have all but rendered Catholic institutions, including hospitals, ineligible for aid. Although Catholic hospitals emphasized that their doors were open to patients of any religious persuasion, they were clearly managed by members of one religious denomination: the female religious communities.

Catholic delegates opposed the amendment, and so did other delegates with more practical concerns. Their opinion, which would surface again and again in the debates that followed, was that the money the state appropriated to these sectarian institutions amounted to substantially less than funds required to maintain public institutions for the same purposes. If the private organizations could not continue their work, public institutions might need to provide more services, and that would be a very expensive proposition indeed.

At the 1867 convention, the issue died when Democrats and Republicans realized that to continue the fight could mean political damage to both parties. However, the issue was soon resurrected when, in 1872, the notorious Tweed ring was voted out of office and a new Republican reform legislature refused to pass the annual charity appropriations bill in 1873. The newly elected legislatures were anxious to change the existing system and established a Constitutional Committee to propose a constitutional amendment prohibiting all public grants to sectarian institutions. The reformers were not entirely successful: a change in New York State policy was implemented in 1874, but the amendment enacted did not go as far as its backers had hoped. It only prohibited some appropriations and still allowed those institutions denied legislative grants to receive public funds through local governments. This change did, however, end direct state appropriations to Catholic hospitals. Still, although the Catholic hospitals in New York City and Brooklyn no longer received direct funding from the legislature, they could receive public aid through local authorities. When, in 1894, a new Constitutional Convention rehashed the debate of 1867, the arguments on both sides remained the same. Critics of the system, for example, objected to what they felt was a high proportion of state funds allotted to Roman Catholic charity institutions.[50]

Indeed, many of the sectarian institutions that received state grants were Roman Catholic. Catholic hospitals throughout the state, including those in New York City and Brooklyn, had fared well under the new laws of 1868. Between 1868 and 1870, the legislature appropriated $288,699.27 to private hospitals. Of the 28 hospitals receiving these monies, 8 were Roman Catholic and received approximately 32 percent of the total funds. However, the percentage of funds granted to Roman Catholic hospitals was actually declining. In 1868, Catholic hospitals received 46 percent of the total grants; in 1869, 34 percent; and in 1870, 26 percent. This decline reflected the growing number of all kinds of private hospitals throughout the state after the war.[51]

From the Catholic point of view, many of the seemingly nonsectarian private institutions receiving funds were clearly Protestant in nature, although not officially affiliated with any one particular Protestant group or church. In this sense, the circumstances were not unlike those surrounding the school controversy in the 1840s, when the Catholic Church protested that the private organization that held a monopoly on state funds for education in New York City was really a Protestant organization. Catholics complained that education in New York public schools was clearly Protestant and often anti-Catholic. Similarly, Catholics complained that private hospitals were often Protestant institutions and that the religious orientation of the hospitals was very apparent and often offensive to Catholic patients.[52]

When the issue was debated again in 1894, opponents of all public aid to sectarian charities once more rallied their forces, but with less than satisfactory results. In a compromise agreed to by the New York archbishop Michael Corrigan, the Catholic Church promised to end agitation for public aid to parochial schools in exchange for continued public support of sectarian charities. The defeat of Democratic candidates in the election for constitutional delegates made compromise a difficult prospect for the church to ignore.[53]

Despite opposition, public aid to Catholic hospitals continued for a number of reasons. The power of the Catholic vote in New York State, particularly New York City, gave the church a significant political voice, but the bottom line was cost efficiency. As the Catholic Church was quick to remind public authorities, it was cheaper to minimally aid sectarian charities than to maintain public ones.[54]

Many public officials also considered this system to be the most efficient way to avoid even greater public expenditures. They recognized that the neediest people were often reluctant to confine themselves to public-run hospitals because of their reputations. If left unattended, these patients would very likely become completely destitute and, ultimately, costly state dependents in the poorhouse. Even reluctant supporters of public aid to Catholic hospitals concluded that those institutions might have a good chance of attending to the needy sick before they were likely to become a permanent public burden. Still, legislators wanted to keep amount granted to Catholic institutions as low as possible. The state's perspective was that "public contributions . . . should be within such limits as will encourage private charity."[55] The New York State Board of Charities, in which Catholics were very deliberately included as members after 1894, kept a lookout for waste and extravagance.[56]

The history of New York City appropriations to Catholic hospitals is as tied to political circumstances as economic ones. Until 1898, New York City appropriations to private hospitals was an unsystematic procedure of flat grants, and the disbursement of funds was part and parcel of the ward boss politics in New York.[57] Other than the promise of political support, there were no strings attached to these grants. The hospitals receiving funds were not required to make any reports or justify how this money was spent. Concluding that this system was inefficient and corrupt, turn-of-the-century reformers instead initiated a per diem, per capita system of reimbursement to hospitals. After 1898, hospitals were not granted funds unless they reported them to the City Charities department and proved that the appropriations were legitimately earned. Hospitals receiving funds were required to submit data indicating patients' social class and medical condition and to submit to inspections by city authorities. They had to show that they were providing city-established levels of hospital care, and the city only reimbursed for patient care that fit its criteria.[58]

In true progressive style, the reformers who initiated this new system were convinced they had maximized efficiency and raised standards. As David Rosner has shown, the progress was not as clearly defined from the perspective of the hospitals receiving the monies. Under the old flat grant system, hospitals did not have to worry about how many patients they treated. Since under the new system they were paid on a per patient basis, smaller hospitals could look forward to receiving less city funding than they had in the past. Furthermore, many hospitals did not conform to the model of hospital care that reformers prescribed. Under the revised system, for example, elderly patients might not all be considered appropriate hospital patients if they were not clearly ill.[59]

Catholics complained about the uncertainty of the public funding system. The percentage of public funds relative to the maintenance of individual Catholic hospitals varied tremendously among hospitals and from year to year (Table 4.3). In 1873, for example, 80 percent of St. Peter's revenue was from public sources while St. Vincent's received less than 8 percent. Still, that figure was high for St. Peter's: in 1870 and 1903, public revenue accounted for 49 percent and 42 percent of its total income.

Overall, public revenue was inconsistent. Public funding for St. Francis' fluctuated tremendously too, ranging from 88 percent of income in 1873 to 2 percent in 1892. There was some consistency, however, as to which hospitals received the most public funding. Year after year the same institutions received large amounts (relative to their expenses) while others continually received proportionately smaller sums. St. Vincent's public funds, for example, were always a significantly lower percentage of total hospital income than St. Francis' or St. Peter's. Given the inconsistencies from year to year and the sporadic nature of allotments, public money never entirely supported a Catholic hospital but was one of several ways it survived.

*Table 4.3* Public Funding of Catholic Hospitals, 1868–1904

| | | *Public Funds* | |
|---|---|---|---|
| *Hospital* | *Year* | *Amount* | *% of Total Income* |
| St. Francis' | 1868 | $3,735.52 | 10.3% |
| St. Francis' | 1870 | 16,263.02 | 26.4 |
| St. Peter's | 1870 | 7,500.00 | 49.0 |
| St. Vincent's | 1870 | 19,168.00 | 44.0 |
| St. Francis' | 1873 | 6,000.00 | 88.2 |
| St. Peter's | 1873 | 5,000.00 | 80.6 |
| St. Vincent's | 1873 | 6,000.00 | 8.1 |
| St. Francis' | 1874 | 3,000.00 | 8.2 |
| St. Vincent's | 1874 | 116,816.24 | 2.1 |
| St. Francis' | 1879 | 3,000.00 | 12.5 |
| St. Vincent's | 1879 | 2,500.00 | 5.5 |
| St. Francis' | 1880 | 5,000.00 | 13.3 |
| St. Vincent's | 1880 | 2,500.00 | 5.3 |
| St. Catherine's | 1892 | 9,789.24 | 13.9 |
| St. Francis' | 1892 | 750.00 | 2.0 |
| Columbus | 1903 | 9,035.00 | 39.0 |
| St. Francis' | 1903 | 20,665.76 | 9.7 |
| St. Catherine's | 1903 | 18,050.70 | 17.7 |
| St. Vincent's | 1903 | 30,909.14 | 12.5 |
| St. Peter's | 1903 | 15,221.88 | 42.2 |
| St. John's | 1903 | 29,257.91 | 58.6 |

*Sources:* New York State Commissioners of Public Charity, *Annual Report 1868,* 216–17; *Annual Report 1869,* 218–20; *Annual Report 1870,* 164–67. New York State Board of Charities, *Annual Report 1873,* 88–93; *Annual Report 1874,* 78–87; *Annual Report 1892,* 560–61. U.S. Department of Commerce, Bureau of the Census, *Benevolent Institutions 1904* (Washington DC: U.S. Government Printing Office, 1905), 180–84.

By the turn of the century a number of New York hospitals that had been founded as charity hospitals began to rely more on revenue from paying patients to meet expenses.[60] Most Catholic hospitals had charged patients a fee from the time they first opened, however. While willing to offer free service to those who could not pay, Catholic hospitals almost always expected patients to make some kind of payment. St. Vincent's, for example, attempted to attract paying patients as soon as it opened. The *Freeman's Journal* described services at the cost of three dollars a week and also noted that private rooms were available. In 1851, another notice for the hospital said, "Patients desiring it can be accommodated with well ventilated and private apartments."[61] A majority of patients treated at St. Vincent's Hospital before the turn of the century paid something toward their care though there were proportionately more paying patients in the hospital's earlier years. In eleven of the years between 1863 and 1900 where such figures are available, about half the patients at St. Vincent's paid something toward their care; at least 30 percent paid in full.[62]

Payment was never a requirement, however, and the procedure was not standardized or regulated. Furthermore, fees varied among Catholic hospitals, reflecting the different patient population at each institution. Sisters determined what their patients would be able to pay based on what they knew about them. In 1897, St. Francis' charged five to ten dollars a week, depending on services and ability to pay; St. Elizabeth's, eight to ten dollars; and St. Vincent's, fifteen dollars a week.[63]

Historians have different interpretations of the trend to charge patients for their hospital care. David Rosner explains how after the changes made in the disbursement of public funds to private hospitals in New York in the 1890s, those of very meager means were required to pay for care they once could receive free.[64] Charles Rosenberg suggests that constant financial worries may actually have actually promoted a comparatively favorable image of Catholic hospitals in the eyes of immigrants. Referring to the "dignity of pay" in Catholic hospitals, he notes how even partial payment might lessen the humiliation attached to the need for hospitalization. Catholic hospital patients who paid something toward the cost of their care could view themselves differently from other hospital patients entirely dependent on the institution's charity. The long-term result was that hospital care lost some of the stigma of pauperization that opened up the possibilities of hospitalization to a broader segment of the population.[65]

Even though many patients at Catholic hospitals were paying for their treatment, named donors still liked to refer to their contributions in terms of charity and stressed a class distinction between patients and donors. At the turn of the century when a majority of patients paid something toward the cost of their care at St. Vincent's and a third paid the entire cost, the Ladies Aid Society still described their work as being "towards the comfort of the destitute sick," and a patron's gift of dinner and fruit for "her ward" in 1906 deliberately conveys an air of benevolent paternalism.[66]

Sisters were raising money in new ways too, for several different, but related, reasons. First of all, sisters recognized that they needed to publicize their hospitals outside the Catholic community to obtain public funds. Fairs continued but not to the same extent that they had in the nineteenth century. While earlier fairs were touted for their financial success, later ones were considered successful public relations efforts. A fair held for St. Vincent's in 1906 was praised because of the money raised but also because it was "distinctly useful in bringing the Hospital prominently before the public."[67]

Second, the Catholic community was changing; it was not exclusively or overwhelmingly poor or even working-class. Fundraising efforts reflected this change. Lay women continued to be involved in hospital fundraising—in fact their numbers grew—but they were organized more formally in hospital auxiliaries, which numbered members in the hundreds by the turn of the century. More genteel events like card parties, teas, and luncheons supplanted the fairs as the primary functions organized by women and reveal the changing class composition among Catholic New Yorkers. An invitation to a tea sponsored by St. Vincent's Ladies Auxiliary at the Waldorf Astoria is a significant change from the boisterous, fun-for-all fair women organized in 1856.[68]

Lay women's hospital work remained gender specific. Publicity surrounding the opening of Mary Immaculate in 1904 noted that its women's auxiliary was organized to raise money for general purposes but also specifically to furnish women's wards for the new hospital. There was another change as well in a stronger emphasis on the religious component of women's participation in hospital fundraising. While hospital administrators had always stressed the benevolent effect of gift-giving on donors, the theme of personal spiritual fulfillment was increasingly incorporated into the activities of the ladies' auxiliaries. St. Vincent's Ladies Auxiliary in 1906, for example, conducted a retreat for members to arouse "the noblest impulses of the soul" and acquire the strength "to attempt better things for God and his Beloved Poor."[69]

In 1905, St. Joseph's Hospital in Far Rockaway opened in startling contrast to the earlier and quieter foundations that reflected the routine, domestic qualities sisters sought to impart to Catholic hospital care when they first began. St. Joseph's was christened with a celebration even before the hospital was ready to receive patients. Music, speakers, and refreshments on the hospital grounds heralded the occasion, and donations were presented and eagerly accepted. In the midst of the festivities a patient presented for admittance was sent off to another hospital for treatment—clearly other requirements had priority that day. This was a public occasion in a way earlier hospital openings were not.[70]

Fundraising efforts and support for St. Joseph's came from beyond the surrounding Roman Catholic community. St. Joseph's supporters continued Catholic hospital tradition and organized a fundraising fair in 1907, but unlike the Crystal Palace Fair organized for St. Vincent's in Manhattan in 1856, most of the booths at St. Joseph's fair were organized by town, not parish. Fairgoers saw town names on banners instead of parish ones—Far Rockaway, Lawrence, Cedarhurst, Woodmere, and Edgemere—and attended a community, rather than a church, event.[71]

St. Joseph's origins differ from other hospitals founded for a particular Catholic group. Still, its history shows how hospital funding had changed. In terms of their finances and fundraising techniques, Catholic hospitals were moving way from their immigrant roots.

Throughout the late nineteenth and early twentieth centuries, individual and group contributions were sporadic, dramatically different in individual amounts, and rarely fixed or guaranteed in any way. At some hospitals, pay patients were an important source of support, at others less so. Public funding varied among hospitals and was particularly crucial to some. Overall, Catholic hospitals survived because of the commitment of the sisters and their remarkable ability to sustain support for their work.

As Catholics liked to point out, even church critics recognized that the sisters did not receive salaries. Almost all literature about Catholic hospitals noted that "Sisters serve for life, with no expense to the Institution save board, the Motherhouse . . . furnishing their apparel."[72] The Catholic press often reminded readers that "sisters receive no compensation" and compared this to the cost of maintaining a staff at other hospitals.[73] *Catholic World,* for example, noted in 1868 that the orderlies at Bellevue Hospital received fourteen dollars a month while Sisters of Charity served for free.[74]

But this focus on the sisters' relatively low labor costs overshadowed the fact that the religious communities that ran hospitals were also responsible for the sisters who staffed it. Although nonsalaried sisters saved a hospital money on its expense sheet, their religious communities supported them and the hospitals out of the same resources, which also went toward other institutions and sisters as well. Other hospitals did not pay much in the way of nursing staff salaries anyway. As noted in chapter 3, until the last part of the century, lay nurses in other hospitals were often patients themselves and were, at best, miserably paid. By 1900, nonsalaried student nurses staffed hospital wards in most hospitals, including Catholic ones.[75]

Hospital sisters were, of course, much more than cheap available labor. Comments that emphasized the sisters' frugality, their service without compensation, or even their saintliness, which put them beyond such mundane concerns as money, failed to recognize the deliberateness of their efforts and the skill and expertise that went into their hospital work, including fundraising. Their financial ability did not surprise those who knew them best. When Sister Mary David, founder and superior at St. John's Hospital, paid off the hospital's mortgage in 1901, the bishop wrote her that bank "officials have asked several times when the hospital was going to draw on them for the rest of the amount the bank agreed to advance" and noted that he himself was "no less surprised to have this indebtedness paid off."[76] His astonishment was probably overstated; he knew how capable the sisters were.

Since little hospital support was ever fixed or guaranteed, hospital sisters always characterized their finances as precarious. The message the sisters sent out in annual reports, in the Catholic press, and throughout the immigrant world of New York was that their hospitals teetered on the edge of financial insolvency. It was an efficacious tactic. While they did struggle at times for lack of money, their pleas solicited responses: the public clearly wanted these hospitals to continue and were willing to help financially, either as contributors or paying patients. Sisters' fundraising reflected enormous skill; it would not be an overstatement to say that it rivaled their nursing as the critical factor in their success. That skill was rarely discussed in the same glowing terms that usually described their nursing—in fact, it was hardly mentioned at all—but that was of no consequence to sisters. Their concerns were about expanding and improving their hospitals.

CHAPTER FIVE

# "Trust in God but Put Your Shoulder to the Wheel"

## *Hospital Sisters and Modernization*

New York's Catholic hospitals entered the twentieth century with their feet in both the old and the new world of hospital care. While general hospitals had fewer chronically ill patients than in prior decades and the number of surgical cases increased, this shift was not complete. At Holy Family Hospital in Brooklyn, the number of operations did not rise significantly until after 1913, increasing from 111 that year to 543 in 1914 and steadily increasing each year thereafter. Alcoholic patients continued to be treated in large numbers at St. Mary's Hospital in Brooklyn. Between 1913 and 1917, St. Mary's treated approximately five hundred patients annually for alcoholism. It was not until after 1919 that this number began to fall.[1]

By then, Catholic hospital sisters were confronting a powerful reform movement in American medicine, one fueled not by any new scientific breakthroughs but rather by contemporary ideas about process and progress. Beginning in the 1890s, reformers—including physicians, philanthropists, nursing leaders, and local government officials—set about to redefine the structure and organization of American hospitals. Applying Progressive era theories of education, efficiency, and standardization to hospital care, their goal was to standardize and modernize the American general hospital: administration, architecture, patient care, and even laundry service were all targeted for reform. They were remarkably successful. By 1929, when more economic concerns took priority, many of their goals had been realized.[2]

Catholic hospital sisters in New York approached the reform movement with the same deliberation that had characterized their earlier efforts to open hospitals, combining religious fervor with an equally strong dose of pragmatism and conformity to mainstream medical practice. As one sister admonished in 1923, "We must not . . . expect

*Fig. 8.* This photo from an annual report for St. John's Hospital Long Island City illustrates something of both the old and new worlds of Catholic hospital care in the early twentieth century. Professional doctors and nurses are ready to work, but look who is overseeing what goes on. Photograph ca. 1908; Sisters of Saint Joseph, Brentwood, New York

God to do everything and we do nothing. It will be well, I think . . . to trust in God, but put your shoulder to the wheel."[3]

Among the issues nursing leaders addressed was one sisters had already confronted: nursing education. As in the wake of the nursing reform movement of the nineteenth century, sisters were again under pressure to meet new professional expectations. This time, their schools, not necessarily the sisters, were the issue. Several of New York's Catholic hospitals had opened training schools by 1910: St. Mary's in 1889, St. Vincent's Manhattan in 1892, St. John's in 1900, and St. Catherine's in 1907. By 1920, nine of the general care hospitals had nurse training programs, and all did by 1930. Hospital sisters in New York accepted most of the goals of the secular proponents of professional nursing and made adjustments and plans to conform to what nursing leaders were anxious to enact, namely more regulation and standardization in nursing education. Despite this, Catholic reaction to the rhetoric of professional nursing still sought to distinguish Catholic nurses' training from that at other hospitals.[4]

Like Catholic women's colleges, the nursing schools were supported by Catholics who feared the effect of secular education and professional training on young Catholic laywomen.[5] Agnes Copeland, supervisor at St. Catherine's Hospital Nurses' Training School and an active proponent for upgrading Catholic schools, explained that it was dangerous for Catholic women to attend nurses' training courses at non-Catholic hospitals because "many of my own acquaintances have lost their religion" in such schools.[6] In 1922, Catholic Charities of the Archdiocese of New York elaborated on the nineteenth-century Catholic position that argued that moral instruction was a fundamental part of all education: "Considering the dangers that surround a life as this, it is difficult to understand why our Catholic girls will select for training a hospital other than our own. Religious influence is an important factor in any line of education and this is particularly true in the training of a nurse in which many moral and ethical problems are involved."[7]

While Catholic nursing schools emphasized how they were different in this one way from other schools, it was also true that nurse training programs at all hospitals duplicated many aspects of convent life. Student nurses were almost without exception required to be single. Nursing school superintendents wanted students free from any other responsibilities or demands on their time. The students' lives, in class and out, were precisely scheduled and monitored by superiors, much like the carefully prescribed religious life. Student nurses followed a highly supervised and rigid schedule very reminiscent of the sisters' schedules.[8]

At one school, for example, students attending Mass were awakened at 5:40 A.M., others at 6:15, but all were required to participate in morning prayers. "On signal at 6:40 morning prayers are held for all nurses regardless of denomination. After morning prayers all repair to the dining room and immediately after breakfast report for duty." Until seven in the evening every hour was supervised and accounted for. Three nights a week nurses attended classes and "All lights are extinguished at ten o'clock except on Saturday and Sunday nights."[9]

As at other hospitals, nursing students became the backbone of the nursing staff in most of New York's Catholic hospitals. As nursing historian Barbara Melosh notes, the student nurse "took her place in a world of female authority" where "superintendents drilled and disciplined her, constantly reminding her of her special duties," much like the young woman entering a religious community.[10]

The proportion of graduate nurses to pupil nurses varied among Catholic hospitals and, again, as among non-Catholic hospitals, reflected the size of the nursing school. In 1920, over one half of the nurses at Misericordia Hospital and St. Vincent's Hospital in Staten Island were graduate nurses, which was unusual in American hospitals in this period. At hospitals with larger training programs, like St. Vincent's in Manhattan, St. Catherine's in Brooklyn, and St. Joseph's in Long Island City, student nurses made up a majority of the nursing staff.[11]

Hospitals had an economic incentive to open nursing schools because more student nurses meant more patients could be admitted. In 1907, for example, St. Vincent's in Manhattan enlarged its training school "due to the increased number of patients."[12] As the New York State Department of Education explained in 1912, for a hospital with limited financial resources, "about the only hope it has of success lies in securing a sufficient number of pupil nurses to enable it to care for the patients at minimum expense for nurses."[13]

This national trend toward a reliance on the labor of nursing students in hospitals was a concern to nursing reformers and a motivating force in their ongoing quest to reorganize the system of nurse training in the United States. The reformers wanted a greater emphasis on the student's education and less on their benefit to the hospital labor pool. Seeking professional recognition and power for nurses, twentieth-century nursing leaders directed their efforts toward expanding the curriculum and initiating state licensing of schools and graduate nurses.[14]

Included in curriculum reform were guidelines requiring that a training school provide theoretical and clinical training in five areas: medicine, surgery, obstetrics, pediatrics, and dietetics.[15] The practical obstetric training was problematic for many New York hospital schools without maternity facilities but not for Catholic nursing schools. Schools at hospitals without maternity departments used the facilities at another hospital run by same sisters. The Sisters of Charity and the Misericorde sisters, in particular, managed maternity hospitals, nursed obstetrical cases, and included obstetrical training in their schools without controversy. St. Vincent's School of Nursing introduced obstetrics in 1894—two years after opening—and students received similar training at the Sisters of Charity's other institutions, St. Mary's Maternity Hospital in Brooklyn and St. Ann's Maternity in Manhattan.[16]

For one group of New York sisters, however, the obstetric requirement presented a different kind of problem. The Franciscan Sisters of the Poor who ran St. Francis' and St. Peter's hospitals were prohibited by the rules of their community from engaging in maternity work. Franciscan sisters therefore could not enroll in a nursing program that included obstetric courses because this kind of work was seen as unsuitable for sisters. Similarly, as Carol Coburn and Martha Smith explain in their study of the Sisters of St. Joseph of Carondolet, early-twentieth-century nursing sisters had opponents within the Vatican who objected to their nursing male patients because they considered it unseemly for "virgins dedicated to God."[17]

Unlike the other sisters with hospitals in New York, the Franciscans had strong connections with a European motherhouse and, arguably, even stronger European traditions, and restrictions. In 1916, the German motherhouse in Aachen advised Cincinnati archbishop Henry Moeller, in whose jurisdiction the congregation's American community originated, that the Franciscan sisters were forbidden to take maternity courses, and pressure to do so might make it impossible for Franciscans to continue with all hospital work. Adamant about her position, the Franciscan superior in Germany explained that "If obstetrical work is absolutely requisite in the course of training for the State examination, the Sisters shall have to restrict themselves to the care of the Poor, Aged and Incurables." Referring specifically to a hospital in Dayton, Sister Hildegard told Moeller that "We shall rather be willing to abandon St. Elizabeth's Hospital entirely to be transferred by Your Grace to some other community than permit the Sisters to assume charge of the maternity wards or take a course in obstetrical work."[18] After Mother Hildegard visited the United States in 1922, the Franciscans changed their policy and, after that, a nurses' training school opened at St. Peter's Hospital in 1923. Franciscan sisters still were given firm rules regarding their own participation in obstetric cases and could only assist in emergencies. The Franciscan's experience was unique among the New York hospital sisterhoods.[19]

Increasingly, nursing sisters headed the nursing programs at their hospitals but the sisters initially had hired laywomen, graduates of non-Catholic training programs, to organize and run their schools. The first director at St. Vincent's School of Nursing in Manhattan was a laywoman, Katherine Sanborne, a graduate of the New York Hospital Training School. In an unusually long term of office, she kept her position at St. Vincent's for forty-two years. By 1930, she was the only laywoman in charge of a Catholic hospital nursing program in New York.[20]

In most cases, sisters ultimately took over the direction of their hospital schools. After Sanborne's retirement in 1934, Sisters of Charity ran St. Vincent's School of Nursing in Manhattan. Some communities continued to hire laywomen, however. In 1932, the Sisters of St. Joseph appointed one of their own members as director of St. John's Long Island City Hospital School of Nursing, Sister Thomas Francis Cushing, but after Cushing left that post to become the general administrator of the hospital, she was replaced with a laywoman.[21]

At St. Catherine's, the earliest directors were laywomen, but after 1922 Sisters of St. Dominic headed the nursing program there. One of the hospital's early lay administrators, Nora McCarthy, left St. Catherine's in 1914 to take over the administration of the nursing school at the Sister of St. Joseph's hospital, St. John's hospital in Long Island City. McCarthy was a graduate of the Sisters of Charity's school at St. Mary's in Brooklyn. Her career suggests how close the world of Catholic nursing was in this period and how it crossed the boundaries of religious communities.[22]

Reform efforts also addressed the education of physicians. Some Catholics also chose to pursue this path of equal yet separate facilities with regard to physician training, but their efforts in New York were not successful. (In the nineteenth century, most physicians trained in privately operated medical schools unaffiliated with either universities or hospitals, and clinical and laboratory work was minimal. Reformers wanted to change that, and did.) New York's only Catholic medical college, at Fordham University, opened in 1905 and closed in 1921.

The medical school was founded when the Jesuits in New York, anxious to propel their school into university status, opened a medical school and a law school. It did not do poorly in the famous Flexner survey of 1910, which rated medical schools on a model with a variety of categories and became a benchmark for the medical school reform movement. In terms of entrance requirements, Fordham's requirements were acceptable, as was its teaching staff. Laboratory facilities were noted as "adequate for the routine of the small student body." Clinical training was available at the nearby Fordham Hospital and Dispensary, but, contrary to Flexner recommendations, the school had no control over the hospital or the staff appointments. Fordham's financial situation also was a problem. Flexner noted that the only monies available to the medical school were the student fees and "appropriations amounting to several thousand dollars annually from the general funds of the university."[23]

Overall, Fordham ranked with other schools that had yet to meet all the Flexner qualifications but were considered to be on the right track. It was not a proprietary school organized primarily to make money, one of the primary violations according to the reformers' standards. Furthermore, unlike some schools that complained about the changes suggested, the college responded favorably to the Flexner survey. Four years later the school fared well in another evaluation conducted by the American Medical Association.[24]

Yet Fordham was never able to fulfill what was to become a major reform requirement in the decade following the report—acquiring a hospital of its own. In 1911 a physician on the medical faculty offered his private maternity hospital to the school, and it applied to the State Board of Charities for reincorporation as Fordham University Hospital. When it was revealed that the physician had been expelled from the American Medical Association, the school asked that its name be removed from the hospital.[25]

There is no indication that Fordham ever attempted to combine with a Catholic hospital to create the institutional complex envisioned by reformers. To outsiders that might have seemed a likely path to pursue, as all were Catholic institutions. In another very fundamental way they were all very separate institutions. The hospitals were maintained by religious communities, not the Roman Catholic Archdiocese of New York or the Diocese of Brooklyn. Similarly, Fordham was not a diocesan institution, it was owned by the Jesuit Fathers. Combining these separate institutions was not as simple as it might have appeared to those unfamiliar with their histories and the details of their organization and management.

Control was a critical factor in negotiations surrounding the merger of a medical school and a hospital. As historian Kenneth Ludmerer explains, many hospital trustees opposed medical school affiliation and many medical schools were unsuccessful finding a hospital willing to accept the reorganization and subsequent loss of autonomy. The Flexner report was certainly very clear about who would be in charge in a medical school union. "Centralized administration of wards, dispensary, laboratories, as organically one, requires that the school relationship be continuous and unhampered . . . The control of the hospital puts another face on its relation to the clinical facility."[26] At the same time, hospital administrators questioned the priorities of a teaching hospital. Patient care did not always seem to be the chief concern at these kinds of institutions. As Jane Addams observed, "the patient's comfort was 'sacrificed to the hospitals looks.'"[27]

Fordham's finances were also a major problem. As early as 1906 the school was operating at a deficit. Unlike the law school, cheaper to run and fortunate enough when it was in financial trouble to find a dean who agreed personally to meet all deficits for five years, the medical school had no luck finding a benefactor.[28] Faced with the possibility of closing in 1919, school officials approached the Archdiocese of New York for financial help, but the new archbishop, John Cardinal Hayes, refused the request. Hayes was more concerned with the need for a Roman Catholic approach to sociology and social work than with medicine. (One of the major efforts of his early administration was to organize the various Catholic charities in the archdiocese under a diocesan umbrella.) He was also probably reluctant to finance the Fordham Medical School because the student body there was not predominantly Catholic. Statistics on the religious composition of the Fordham Medical School are not available, but clearly all the students were not Catholic. Commencement procedures even made allowances for other students—a Jewish graduate remembered that he was exempted from kissing the cardinal's ring during the ceremony.[29] The school's demise suggests that the concept of Catholic superiority and expertise in the care of the sick that was so often raised in support of Catholic hospitals in the nineteenth-century did not extend to a recognition of the need for a Catholic insight or perspective on modern medical education in New York.

Another reform emphasis involved efforts to upgrade (in the reform language) the requirements necessary for hospital administrators. This had always been a position that sisters in Catholic hospitals held. Some hospital reformers argued that the duties of a hospital administrator were more complicated than in the past and, as a result, required special training. They claimed that because acute care translated into a greater turnover in patient population, because technological changes and growth made for a more complex physical structure, and because rising costs demanded a more elaborate financial structure, hospital administrators needed special skills and experience in administration.

The argument undermined the traditional characterization of the job as one requiring feminine qualities. Some reformers considered the sisters to be unsuitable administrators by simple reason of who they were.[30] A survey conducted by the Brooklyn Diocese of Catholic hospitals in Brooklyn and Queens in 1923, for example, noted with disapproval that the superintendent at St. Joseph's Hospital in Far Rockaway was not sufficiently trained for hospital administration. She was not a nurse and although she had executive experience in schools, until her appointment to St. Joseph's she had never worked in a hospital.[31]

The emphasis on formal education for hospital administrators remained more talk than action, however, in both the secular and Catholic hospital world, although specialized training was available through individual courses, often in postgraduate nursing education.[32] In New York, the graduate program in nursing at Columbia University included a course in hospital administration. A sister from St. Mary's Hospital in Rochester, Minnesota, (associated with the Mayo Brothers clinic) who attended this program in the early 1920s encouraged other sisters to attend, but she pointed out the difficulties as well. She noted that courses and field work were excellent and that daily contact with all different people, especially nonreligious, was "broadening," but she also complained that it was physically difficult for sisters to attend Columbia. Accommodations were limited, and sisters missed their community life. For the time being, graduate work in hospital administration remained an ideal rather than a necessity.[33]

When they could, New York's Catholic hospital sisters conformed to the administrative and organizational aspects of the reform movement. Sister Marie Immaculate Conception of Misericordia Hospital supported a new criteria to choose hospital supervisors, explaining how the "sister-nurse, carefully and efficiently trained as she may be, is not yet prepared to fit into the various executive positions of the hospital. She may be an excellent nurse, unsurpassed in the care of her patients; yet as a supervisor, as a superintendent, she may be a complete failure."[34] Included in the accomplishments profiled in *Hospital Progress,* the journal of the Catholic Hospital Association founded in 1916 by the Sisters of St. Joseph of Carondolet, Wisconsin, and Rev. Charles B. Moulinier, S.J., of the medical college of Marquette University to encourage reform among hospital sisters, were examples of New York success stories. St. Catherine's in Brooklyn, for example, boasted of its accommodations to standardization with extensively detailed reports from staff members: Sister Ildephonse reported on "The Record Room," Dr. DeCoste on "Pathology and X-Ray Labs," and Dr. Gordon on "Obstetrics in St. Catherine's Hospital." Other articles about New York's hospitals reflected the long reach of reform. St. John's in Long Island City was cited for its social service department and new nurses' residence, Mary Immaculate in Jamaica for its fire prevention.[35]

The greatest concern to sisters was the cost of reform. Hospitals had different support networks and some were able to spend more than others. Commenting on the high cost of modernization in 1922, Catholic Charities of the Archdiocese of New York noted that while St. Vincent's in New York was able to spend "$161,000.00 on the improvement of its plant and in scientific equipment . . . Other hospitals in need of improvement were not as well able to meet the financial burden."[36]

Also contributing to financial problems was the fact that many patients did not pay for their total care at Catholic hospitals. That had always been the case, but part of what made Catholic hospitals different from other hospitals was that some patients did pay at least something toward their care. Now patient payments were expected at almost all hospitals and (fundraising appeals from Catholic hospitals frequently noted the number of non-paying patients. At St. Vincent's in 1925, for example, 22 percent of patients were treated free of charge and 26 percent made partial payment. Such was the case at other Catholic hospitals as well. That same year Catholic Charities of New York found that 23 percent of all patients treated at Catholic hospitals in the New York Archdiocese were treated free of charge and 28 percent made partial payment. Another 10 percent were public charges. While the hospitals received reimbursement from the city for these patients, according to Catholic Charities, the reimbursement did not cover the actual costs of care.[37]

In the Brooklyn Diocese some Catholic hospitals treated a greater percentage of free patients than others, again an indication of the distinctions between Catholic hospitals. In 1922 almost all patients at St. Joseph's Hospital paid for their care (94%), yet over one-half of the cases treated at St. Peter's were free patients. The other hospitals in the Brooklyn Diocese all had over 50 percent paying patients: Mary Immaculate, 76 percent; St. Catherine's, 69 percent; St. John's, 68 percent; St. Mary's, 66 percent; Holy Family, 64 percent. City charity cases were highest at Holy Family (29%) and St. John's (19%), while St. Peter's did not have any.[38]

Financial restrictions limited the extent to which sisters could upgrade to meet reform standards. In 1921 the Sisters of Charity sold their newest hospital, St. Lawrence, because it needed improvements that the Sisters of Charity could not afford. Concluding that the community needed an expanded facility as soon as possible, they sold it to the Missionary Sisters of the Sacred Heart, who promised to develop it immediately. The proceeds from the sale of St. Lawrence were disbursed among other institutions the Sisters of Charity owned, including St. Vincent's in Manhattan. Facing similar circumstances, when the Sisters of St. Dominic made plans to expand and upgrade Mary Immaculate Hospital, they transferred the ownership of the property to the Diocese of Brooklyn and converted the corporation to a diocesan hospital. In doing so the hospital became eligible for increased diocesan financial assistance.[39]

Fundraising continued to be a major concern of Catholic hospitals in the 1920s. Fundraising efforts at Mary Immaculate in Queens and St. Vincent's in Manhattan, both involved in major rebuilding in the 1920s, reveal how reform methods extended to fundraising and that Catholic hospitals recognized a need to extend their reach in this new era. At Mary Immaculate, businessmen were brought in to direct the appeal, and two years before the campaign actually began a press agent was hired to publicize the hospital and prepare the community for the need to contribute. The plan was to canvass forty thousand families whose names and addresses had been garnered from church membership rolls, telephone books, and the motor vehicle department. An instruction manual with pertinent information and tactics "made efficient salesmen and saleswomen of the workers" who numbered over 1,500, referred to as an "army" by the chairman of the organizing committee. The appeal was highly publicized: print and radio ads, flyers, posters, mailings, and an essay contest in parochial and public schools "so that all the children . . . might be interested in the work and in turn interest their parents."[40]

The campaign was delayed until an appeal at the nearby Jamaica Hospital had concluded, indicating that the fundraisers recognized that theirs was not a Catholic project exclusively. The campaign emphasized that Mary Immaculate was a hospital serving the entire community. Although the appeal was organized through local Queens parishes, non-Catholics were recruited as workers and for supervisory positions. Unlike nineteenth-century fundraising, which emphasized the need for the continued charity of financial supporters, the theme of this campaign was the community's responsibility for its hospital. Mary Immaculate was characterized not as a charitable institution but as a community service for all citizens on par with the fire department or the police. Campaign literature noted that "The average person will need a hospital 75 times oftener than he will need a fire engine."[41]

Like efforts at Mary Immaculate, St. Vincent's appeal for funds in 1926 was highly organized and directed toward a wide audience. New York mayor Jimmy Walker, once a patient at St. Vincent's, was just one of the donors profiled in a campaign organized to raise funds "to increase its accident and emergency facilities, to provide more treatment rooms for all kinds of cases, to provide five times its present number of beds for children, to provide a maternity department, to pay for the nurses' home now under construction, to enlarge the laboratory, to accommodate more interns, and to supply much-needed x-ray equipment." St. Vincent's appeal was also traditional in tone and stressed the unreimbursed patient care the hospital delivered, as well as citing rising expenses due to modernization. Fundraising literature reminded would-be donors that "in modern society one is unable to render personal service to the sick and needy as the Good Samaritan did, but that institutions like St. Vincent's Hospital can perform the service if the would-be Good Samaritan will supply the means."[42]

Throughout this period, sisters continued to be ultimately responsible for the financial maintenance of their hospitals. Although health divisions were included in the Catholic Charity organizations of both the Archdiocese of New York and the Diocese of Brooklyn, the purpose of these departments was advisory and they were in no way meant to assume financial obligation on the part of the hierarchy for hospital or patient costs. Catholic Charities of the Archdiocese of New York made that very clear in its annual report of 1925, which explained that the "Central Organization [of Catholic Charities] never assumes financial responsibility for the care of any patient in one of our own Hospitals, although we do refer many cases for treatment."[43]

Catholic Charities' interest was in maintaining standards. The same report made clear that each hospital was required to make some provision for the care of the poor, but it emphasized that its financial assistance to a hospital was "granted when it is needed in order to maintain high standards of efficiency, rather than in consideration of the number of charitable cases it receives." Catholic Charities promised that if the hospitals provided care and treatment to the free as well as the pay patients the organization referred, the Archdiocese would "help it meet the requirements which modern standards demand."[44]

Even with those qualifications, that financial assistance was limited. Between 1920 and 1930, Catholic Charities of the Archdiocese of New York only allocated a small percentage of its total appropriations to hospitals. There were several initial appropriations to general hospitals, including ten thousand dollars to St. Vincent's Manhattan and twenty-five thousand to St. Vincent's in Staten Island, and funds allocated to health-related institutions and organizations totaled almost one-third of the money spent between 1920 and 1921. However, after that initial expenditure, Health Division appropriations amounted to less than 10 percent of the Catholic Charities' annual total. Furthermore, allotments in the Health Division were usually for chronic rather than acute care institutions. Convalescent homes, visiting nurse services, mental health, and social service organizations, as well as institutions for the aged and chronic patients, were more likely to receive assistance than hospitals.[45]

The directors of Catholic Charities felt their money was better spent in a chronic rather than acute care hospital. Noting that general hospital care was primarily acute care now, the diocese maintained that it was no longer necessary to provide that care in a Catholic environment. Catholic Charities argued that it was impossible to care for all Catholic patients in a Catholic hospital (pointing out that it was often more convenient for a patient to attend a non-Catholic institution) and concluded that "in individual cases, however, there is reason for insisting on care in one of our own institutions . . . In particular this is true in the cases of Tuberculosis and Cancer. These patients must of necessity spend a long time in the hospital; for them the spiritual comforts and consolations of the Catholic hospitals mean much."[46]

Maternity patients were also a special concern to Catholic Charities, which noted that "it is often highly desirable and many times absolutely necessary that they be cared for in our own hospitals. In these days of Birth Control propaganda we must have adequate accommodations in Catholic Maternity hospitals not only for the poor but for persons of moderate means."[47]

By 1930, much of the old rhetoric used to describe Catholic hospitals was no longer appropriate. One hospital superior expressed concern over continued reference to her institution as a "free public hospital." Her lawyer reassured her that some charity patients were enough to satisfy the corporate definition, noting she "need have no qualms of conscience about characterizing the institution as a free public hospital. [T]he receipt of money from patients who can pay does not in any way detract from the status of a hospital as a free institution if the main purpose of the hospital is charitable, and if this is not a money making institution."[48]

Other points, particularly those that originally had been raised to justify the need for specifically Catholic hospitals, were less easily addressed. The nineteenth-century hierarchy, for example, spoke of the need for Catholic hospitals because they worried about the availability of Catholic services and religious ritual to patients in public and other private hospitals. That was certainly no longer a problem in twentieth-century New York. As Catholic Charities in the New York Archdiocese explained in 1928, "the priest and any other minister of religion is given every courtesy in most hospitals."[49]

A twentieth-century characterization of the hospital as a community institution geographically defined, made older arguments surrounding the need for Catholic hospitals irrelevant. A Brooklyn diocesan survey found in 1922 that at most hospitals a majority of patients remained at least nominally Roman Catholic but a significant number of patients were not Catholic. At St. Catherine's, St. Mary's, and Mary Immaculate, at least one-third of patients specified their religion as either Protestant or Jewish. St. Joseph's Hospital in Far Rockaway had an exceptionally high non-Catholic population. While one-half of the patients at St. Joseph's were Catholic, Protestant, and Jewish patients each accounted for one-quarter. St. Joseph's appeal letter in 1926 was sent to all clerics in the area—Catholic, Protestant, and Jewish.[50]

By the same token, a number of New York City private hospitals that were not Catholic institutions treated a large amount of Catholic patients. Roman Catholics accounted for over one-half of the patients treated at Knickerbocker Hospital and the New York Hospital in 1925. In Brooklyn, 45 percent of the patients treated at Brooklyn Hospital were Roman Catholic, and over a third at Wyckoff Heights and Norwegian Lutheran.[51]

Many patients appeared to be choosing a hospital based on its location rather than its religious affiliation. St. Vincent's still attracted patients from all over the city, so religion continued to be an operative factor in patients' choice there, but at least half the patients in other Catholic hospitals in Brooklyn and Queens lived in the area surrounding the hospital: over 70 percent at Holy Family, St. Catherine's, St. John's, and St. Mary's, and 80 percent at Mary Immaculate. The large Catholic populations reflected a location with a high percentage of Catholic residents, not necessarily the patients' choice to be treated in a Catholic institution. St. Joseph's larger percentage (40 percent) from outside the borough reflects the hospital's location on the border of New York City—areas nearby St. Joseph's included towns in Nassau County.[52]

These changes complicated attempts to delineate a uniquely Catholic responsibility in health care. A priest reminded the International Guild of Catholic Nurses in 1924 that a Catholic nurse should know how to assist a patient to make a perfect act of contrition. But he also acknowledged the difficulty of the task if in fact "the patient has never had the faith."[53]

As the Great Depression worsened, all of New York's hospitals, including the Catholic ones, suffered a loss in patient revenue. The number of ward patients rose, and private rooms, the source of the greatest potential revenue, went unused. A survey conducted in 1933 found that among all nonprofit hospitals in New York, ward occupancy ran at 81 percent, semiprivate rooms at 55 percent, and private rooms at 35 percent. As Rosemary Stevens explains, hospitals were once again increasingly charity institutions but now served a much broader section of the population. Patients who might have been able to pay for hospital care in the 1920s no longer could; even fewer patients could pay for their total care or make partial payment. In 1931, 35 percent of patients treated at Catholic hospitals in the New York Archdiocese paid for their care, 24 percent were treated free of charge, 16 percent made partial payment, and 23 percent were welfare cases.[54]

New York's Catholic hospitals were not alone in their financial problems: the cost of expansion and upgrading was nondenominational, and now the cost of modernization loomed alongside new financial concerns. With the onset of the Great Depression, all American hospitals faced a new and severe economic crisis. Amid that, Catholic hospitals faced an additional crisis of definition.

EPILOGUE

# "A Service So Dear"

In the decades following the standardization movement, an alternative model of hospital care was no longer anything to boast about. Hospitals needed to show the public that they were up-to-date, not different. Moreover, Catholics no longer faced discrimination at other hospitals. As the annual report of Catholic Charities of the Archdiocese of New York noted in 1928, "the priest and any other minister of religion is given every courtesy in most hospitals." The same report explained that a "majority of patients are satisfied with a hospital in which they can receive adequate care for their physical ailments," suggesting some concern within the church hierarchy that a hospital's religious affiliation was no longer of great concern to patients.[1] As a result, Catholic hospital promoters began to characterize Catholic hospitals in a new way, as community institutions and stressing geography rather than religion or ethnicity.

Did Catholic hospitals cease to be distinctive in the early twentieth century and beyond? In one respect they did. Through the 1960s most Catholic hospitals continued to be run by the same communities of women religious that had founded them, and sisters remained in hospital work in a number of ways. They still nursed (even as paid nurses became more commonplace and outnumbered nursing students as staff in the period following World War II); they worked as nursing teachers, pharmacists, x-ray technicians, and the like; and they continued as administrators and as members of hospital boards of trustees. As women, and religious women at that, they were an anomaly among other hospital executives. Photographs show hospital sisters in their traditional habits alongside laymen and women in modern dress and even with other hospital personnel in lab coats or nurses' uniforms, looking like as if they had landed from another world. In some ways they had.

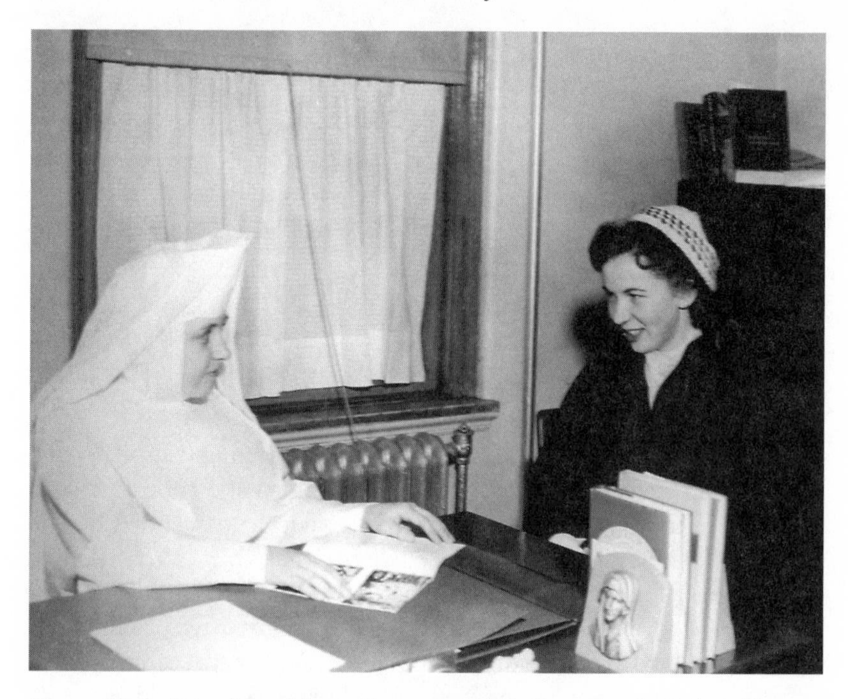

*Fig. 9.* Sister Saint Magdalene Gorman, C.S.J., served at St. John's in a variety of capacities, including as director of nursing from 1946 to 1970, when she left to work at another of the community's hospitals. Her tenure at St. John's is a personal example of a community's longstanding involvement and identification with an institution. Photograph ca. 1955; Sisters of Saint Joseph, Brentwood, New York

For these women, their mission was the same as it had been when they began. As they went about their hospital work, sisters had more on their minds than physically attending to illness. For them, hospitals were part of a wide-ranging agenda, one that included a multitude of other work. Sisters' definition of their mission of charity went far beyond financial matters; their responsibility was to those in need physically and spiritually. For sisters, hospital care was by definition a part of their charitable mission and neither payment nor "modernization" had anything to do with that.

While sisters remained constant in their mission, what did change was the fundamental characterization of Catholic hospitals as sisters' hospitals. Catholic hospital boosters no longer emphasized that their institutions were distinct because the sisters brought something special and superior. No longer did a Sister of Charity's description of her community's involvement at Holy Family Hospital as fundamentally a history of "a service so dear to us" count as the most critical factor in defining superior hospital care.[2]

Like Catholic hospital development in the nineteenth century, the history of Catholic hospitals in the twentieth century parallels some major national trends—it is a story of hospital closings and mergers, not new foundations. The year 1965 marked the beginning of a decline in the number of American hospitals, particularly in urban areas, and the greatest proportion of hospital closings were among the private, not-for-profit institutions. Between 1965 and 1975, the number of Catholic hospitals nationally fell from 803 to 671.[3] In New York, some hospitals closed; in other instances founding communities withdrew and ceded ownership and control to Diocesesan authorities, and some hospitals were merged into Diocesan organizations. St. Vincent's in Manhattan, the city's very first Catholic hospital, was reorganized in the 1980s under the joint sponsorship of the Sisters of Charity and the Archdiocese of New York. In 1990, for the first time in its 140 year history, the director of St. Vincent's Hospital was not a Sister of Charity.[4]

A frequently cited explanation for these changes in Catholic health care is an enormous decline in the numbers of Catholic women religious. As with most assumptions about sisters, it is only partially true. The number of sisters did drop dramatically in the last quarter of the twentieth century, as fewer women joined religious orders and many professed sisters left their religious communities.[5] Most of the reorganization of New York's Catholic hospitals was simultaneous with this change, but the roots of it are deeper and located in the early twentieth century. The circumstances surrounding the Sisters of Charity's departure from two hospitals in Brooklyn, in 1941 and in 1955—decades before the lessening number of sisters—suggest that there is more to the story than a shortage of nuns.

The first hospital the New York Sisters of Charity left was St. Mary's in Brooklyn. As at other Catholic hospitals, the hierarchy's role in the management of St. Mary's was originally minimal. Catholic Charities made recommendations about modernization and standardization, but its role was advisory.[6] When they left St. Mary's, the sisters explained their departure as a result of the pressures of reform and changing expectations. Specifically, the need for increased education for the sisters inhibited their ability to function effectively. Their departure from St. Mary's was ostensibly necessary and agreed on by all, and they continued with their other hospital work in Brooklyn at Holy Family Hospital. The story as they presented it made some sense. The hospi-

*Fig. 10.* Sister Loretto Bernard Beagan, S.C., was a nurse at Saint Vincent's Hospital in Manhattan from 1926 to 1936. She returned as administrator in 1948 and left in 1960 to assume the position of mother general of the Sisters of Charity of New York. Photograph ca. 1960; Sisters of Charity of New York

tal was in weak financial condition, the result of the lingering costs of standardization coupled with the stress of a decade-long economic depression.[7]

Furthermore, St. Mary's had always had more of a diocesan connection than other hospitals. It was founded on the bishop's directive with the proceeds of a diocesan fair. The first president of the board of trustees of the hospital was Bishop John Loughlin, and all consecutive bishops of the Diocese of Brooklyn continued to hold that position. By 1940, only one Sister of Charity remained on the board. In a letter from the board of trustees to the Sisters of Charity on their departure, Vice President Edward Hoar referred to the community's "association" with the hospital, noting that they had been "in charge" for fifty-eight years. There was no sense that this hospital belonged to the Sisters of Charity in any way.[8]

A very different set of circumstances surrounded the community's departure from Holy Family Hospital in 1955. The Sisters of Charity left Holy Family only after unsuccessfully resisting diocesan plans. The controversy between the sisters and the hierarchy began in 1955, when the Division of Health and Hospitals of Catholic Charities of the Diocese of Brooklyn expressed a desire to merge Holy Family with the nearby St. Charles Orthopedic Hospital, which was run by the Daughters of Wisdom. Catholic Charities' plan was framed around the premise that this merger would maximize resources. They concluded that recent medical advancements in polio treatment and rehabilitation lessened the need for St. Charles, a children's orthopedic hospital, and they intended to erect one larger general hospital on the site of Holy Family Hospital. The Sisters of Charity were asked to assume control of the merged institutions.[9]

Advised by the diocese of its plan, the governing board of Sisters of Charity met at their motherhouse and decided that they would not be able to assume responsibility for a larger institution and that they preferred to leave matters the way they stood. The diocese would not take no for an answer and continued to pursue the merger. The future of the plan came to rest on the question of who owned the hospital and could decide its future. Did the Sisters of Charity own Holy Family or did the Diocese of Brooklyn?[10]

The Sisters of Charity thought they did but were advised by the diocese to recheck their records. Upon doing so, they found that earlier in the century this same question of ownership of Holy Family Hospital had been raised but toward a different purpose. In 1925, anxious to receive diocesan endorsement of loans much needed to finance building and renovation (they were in danger of losing their nursing school accreditation), the Sisters of Charity sought to assure the hierarchy that they had no claims on ownership. Mother Vincentia wrote the bishop that although "the incorporators at the request of the late Bishop McDonnell were Sisters of Charity, and they have since 1909 continued to act as managers and trustees, they have never regarded the institution otherwise than as a diocesan hospital and wish in all things, to carry out the intentions of the Bishop and his consultors in their management of the Hospital." While it may have appeared that "the Hospital of the Holy Family is a Community owned hospital," that was not so.[11]

The matter did not end there; two years later the question was still being pursued. In February 1927, the Sisters of Charity were advised that the bishop wished to "establish beyond a doubt, the claim of the diocese to the property, before assuming liability for a large loan for building purposes." To facilitate these arrangements, the bishop insisted that "all papers indicating that the Hospital of the Holy Family is a Diocesan Institution be sent to the Diocesan Attorney . . . all deeds for properties recently purchased with Diocesan funds be also forwarded . . . and hereafter any property that might be bought or any business that might be transacted by means of Diocesan funds and all legal business pertaining thereto, will be settled by [our] Diocesan Attorney." The sisters complied.[12]

Reviewing all these transactions in 1955, the Sisters of Charity had to conclude that the Diocese held the trump card. Agreeing that they were unable to staff an enlarged hospital, their council discussed other options and decided that they would withdraw from Holy Family rather than have it be said that they were forced to leave. When the Diocese of Brooklyn failed to respond to one final assertion of their position on the matter, they announced their willingness to leave at the bishop's convenience. They did so in November.[13]

The paper trail leading to the diocese's ultimate direction of the future of Holy Family Hospital illustrates the constant financial burdens of hospital reform, circumstances certainly not unique to Catholic hospitals. In 1955, the hospital was in desperate need of renovations, and the sisters themselves saw the circumstances as critical. At a general meeting of the community's advisory council, Sister Loretto Bernard, superior at St. Vincent's Hospital, advised the other Council members that "she would not be in favor of remaining [at Holy Family] unless we could improve the quality of care given by the hospital . . . it is not bringing credit on our community or the Diocese of Brooklyn. Accreditation has been withdrawn, there are long lists of violations which could be eliminated only with great effort and vast organization and physical improvements."[14]

While the Sisters of Charity recognized that problems existed and that they were grave ones, not all saw the diocesan plan as the only solution. As her earlier comment indicates, Loretto Bernard was concerned with the level and quality of care at Holy Family, but she suggested another alternative. She recommended coordinating "all community hospital work with the direct supervision of major superiors, to see that standards are conformed to."[15]

These events, and to a lesser extent the earlier ones at St. Mary's, occurred before the enormous decline in the number of women in religious orders, which began in the late 1960s. The post–World War II years saw tremendous enrollment for most female religious orders. In 1951, eighty-three women entered the Sisters of St. Dominic's Novitiate, double the usual number. Among the Sisters of Charity, where entrance numbers had been on a slow decline since the 1920s, there was also an upsurge in membership.[16]

Like others throughout the country, New York communities enthusiastically readied themselves for all the real and anticipated recruits. The Dominicans modernized and expanded their Motherhouse and Novitiate in Amityville, Long Island; the Sisters of St. Joseph did the same at their properties in Brentwood. Both facilities resembled one described by a sister from another community. "The building was huge. Three wings stretched out from the central section, which contained the main entrance, the chapel, and the visiting parlor. One wing was for the professed sisters, one for the postulants, and one for the novices . . . The grounds seemed endless, green and rolling."[17] Their expansion and optimism was part of worldwide sentiment. In the early 1960s, it seemed to many observers in and outside the church that the time had come for a different kind of Catholicism, one more open and active and less authoritative. Pope John XXIII's Vatican Council, which ended in 1965, heralded the way for change.[18]

An important part of the council's message was that there needed to be a larger role for the laity, both male and female, in church activities, but this message was in no way intended to diminish the need for female religious orders. In fact, just the reverse was expected: many sisters and would-be sisters hoped to see that role expanded. Statistics show many women beginning to leave the convent after 1963, but at the same time, many young Catholic women were optimistic about change in their church and, as a result, they continued to enter religious communities.[19] As one former sister who entered a religious order in 1967 recalls, "When I entered the convent it was not as archaic a decision as it might now seem. Convents all over the country were expanding. The tide would very soon begin to run the other way, but in 1966 there were 181,421 nuns in the United States, the most there would ever be. Change was in the air—positive change—and the Catholic Church was part of it."[20]

When reforms were not forthcoming, fewer women entered while more sisters left, and overall numbers declined rapidly. The reevaluation of Catholic attitudes about who was most qualified to run hospitals began before the religious communities felt the numbers crunch, however. At St. Mary's and Holy Family, founding sisters left hospitals for reasons not related to a decline in the number of sisters. An emphasis on the sisters' numbers—or lack of them—perpetuates the idea that Catholic hospitals survived and succeeded primarily because of the cheap labor of the sisters. That sisters did not receive wages in the nineteenth century was not the pivotal factor in their success. New York's Catholic hospitals succeeded because of the way they did their work and attracted patients and financial supporters.

The declining emphasis on sisters as delineators of Catholic hospital distinctiveness in the early twentieth century was concurrent with a change in attitudes within the church about the role and capability of sisters. A codification of canon law in 1918 tightened regulations for sisters and emphasized the separation of their lives from the rest of the world even for communities actively engaged in work outside of their convents. Referring to this "cloistered mentality," Mary Ewens explains how after 1920, "sisters were warned to restrict contact with the outside world as much as possible. Newspapers, radios, libraries and so on, were seen as dangerous distractions, as were various kinds of public events and meetings."[21]

It is not difficult to see how a new perception about the proper role and behavior for sisters would influence attitudes about their capabilities. Apparent in the pattern of events at Holy Family is an implication that the sisters were naive about the real world of hospital management. Catholic Charities applied for the approval of the Hospital Council of Greater New York even before presenting their plan to merge Holy Family with another hospital to the Sisters of Charity. The Sisters of Charity were never asked to participate in the decision-making process—they were given the choice to be part of the new venture (at great cost to their community) or not. Some sisters had a less deferential view of their relationship with the hierarchy. Writing to her bishop in 1925, a superior at Holy Family noted that while "Our intention and desire in the management of the hospital, and our interest in it, have never been other than to serve the Diocese . . . if there is a legal question to be decided we would like to be consulted and represented."[22]

In the context of these changes, some Catholic hospital supporters looked to characterize the Catholic physician as the meaningful factor that distinguished Catholic hospital care from other hospital care. This took some doing since most physicians at Catholic hospitals did not receive any special Catholic medical training. Unlike the numerous Catholic nursing schools, there were relatively few Catholic medical schools and, after Fordham University closed its school of medicine in 1921, none in the New York area until Seton Hall in New Jersey opened a medical college in 1956.[23]

In the 1950s, the Association of Catholic Physicians, a group first organized in 1931, revived an early-twentieth-century point of view, articulated by Rev. Thomas Conaty, rector at Catholic University, about the need for a specifically Catholic medical education to foster Catholic principles and ethics. The association's journal, *The Linacre,* included frequent articles discussing the appropriate training for the Catholic doctor. The dean of the Creighton University School of Medicine in Minnesota, for example, explained the need for Catholic physicians in light of "our present day civilization, with its rank materialism and utter disregard of all things spiritual." In addition to his clinical responsibilities, the Catholic physician, especially the general practitioner, had another "heavy burden" because of his "special role as family doctor and counselor." As a result, "Next to the parish priest the family doctor should gain and hold the confidence of the members of his community. His mode of living and moral standards must be of the highest if he is to keep faith with the trust they place in him." The editorial concluded by noting that the training for this tremendous responsibility was best found in a Catholic medical school.[24]

Such writings likened becoming a Catholic physician to a religious experience as much as a medical one, and they echo earlier remarks about sisters and health care. Nineteenth-century Catholics claimed their hospitals were different because the sisters were religious women and, as such, cared in a special way about their patients. However, physicians were secular men—possibly anxious to bring another level to their clinical work—but very much unlike the sisters whose entire life was organized around religious concerns. Furthermore, not all physicians in Catholic hospitals were Roman Catholics; it was not a requirement for staff positions. Ultimately, Catholic physicians would not distinguish Catholic hospitals the same way the sisters had.[25]

Sisters were never mentioned in this discussion of Catholic doctors because they were never physicians in Catholic hospitals. None of the communities involved in hospital work in New York had ever trained any of their sisters to do so. To begin with, sisters would have faced extreme difficulties pursuing medical education. Although some medical schools began to open to women students in the nineteenth century, the twentieth-century hospital reform movement closed many of the few schools traditionally available to women. Although official papal permission was granted in 1936 for sisters to study medicine, few did. In New York, Catholic sisters were always nurses not doctors. Meeting nursing educational requirements was relatively simple for sisters. Nurses' training, did not challenge any time-held sisterly traditions, was inexpensive compared with the cost of physician education, and was conducted within the sisters' own world, initially in hospitals and, later, at sisters' colleges.[26]

Sisters continued to work and manage Catholic hospitals in New York after modernization, but they were no longer seen as the experts in hospital care exclusively by nature of their lives and identification as religious women. Now sisters worked in their hospitals as trained professionals: as nurses, administrators, in labs. They remained religious women, and, for all that otherworldliness some of their critics found troublesome, New York's hospital sisters had accomplished quite a bit.

Their hospitals that remain are a quiet reminder of a different time for New York Catholics and their church. Sisters were not necessarily saints, or even ahead of their times in matters like class and race which we now recognize as inherently linked to issues of health care and social welfare. While they promised equal care to all, sisters were really only concerned with taking care of "their own." That they did, and their efforts had repercussions for both the city and the church.[27]

Hospital sisters had eased the burdens of illness for several generations of immigrant New Yorkers and left the Catholic Church firmly entrenched in the hospital landscape of New York City. They did so as religious women cohabiting their Godly world and immigrant New York; in fact, they integrated the two easily.

Sisters' faith in the omniscient power of God included their belief, and no doubt many times their hope, that He would heal the sick in their care. Nevertheless, theirs was not a treatment infused with a zealot's dramatic display of religiosity but one that recognized that good health care was more than medicine and surgery. At the same time, the therapeutics of their health care was decidedly noncontroversial and up to date scientifically.

The history of what sisters did in their hospitals in New York City counters some powerful perceptions about nuns which have seeped into our popular culture—contradictory images of menacing psychotics and passive church mice. That history tells a different story, one of determined and pragmatic women who, in choosing an alternative lifestyle for themselves, also embraced the world around them.

# Notes

| | |
|---|---|
| AANY | Archives of the Archdiocese of New York |
| BHA | Bellevue Hospital Archives |
| CUA | Columbia University Archives |
| FSP | Franciscan Sisters of the Poor Archives |
| MSSH | Missionary Sisters of the Sacred Heart Archives |
| NYAM | New York Academy of Medicine Archives |
| NYCMA | New York City Municipal Archives |
| NYH | New York Hospital Archives |
| SJNY | New York Provincial of the Society of Jesus Archives |
| RPA | Redemptorist Provincial Archives New York |
| SCMSV | Archives of the Sisters of Charity Mount St. Vincent |
| SSDCHC | Sisters of St. Dominic Congregation of the Holy Cross Archives |
| SSJB | Sisters of St. Joseph Brentwood Archives |

## Preface

1. Whitfield J. Bell, "Medicine in Boston and Philadelphia: Comparisons and Contrasts, 1750–1820," 163–64; Philip Cash, "The Professionalization of Boston Medicine, 1760–1803," 73–74, 93; and G. B. Warden, "The Medical Profession in Colonial Boston," 154, in *The Colonial Society of Massachusetts: Medicine in Colonial Massachusetts, 1620–1820* (Boston: Colonial Society of Massachusetts; Charlottesville: University Press of Virginia, 1980); James Cassedy, *Medicine in America* (Baltimore: Johns Hopkins University Press, 1991), 6–7, 19–20; Charles E. Rosenberg, *The Care of Strangers: The Rise of America's Hospital System* (New York: Basic Books, 1987), 18–22.

2. J. M. Toner, "Statistics of Regular Medical Associations and Hospitals of the United States," *Transactions* 24 (1873): 314; Rosemary Stevens, *In Sickness*

*and in Wealth: American Hospitals in the Twentieth Century* (New York: Basic Books, 1989), 20, 368n10; Rosenberg, *Care of Strangers,* 1–22.

3. Rosenberg, *Care of Strangers*; Stevens, *In Sickness and in Wealth*. Studies of individual cities and localities include Elizabeth Long and Janet Golden, eds., *The American General Hospital: Communities and Social Context* (Ithaca: Cornell University Press, 1989); Sandra Opdycke, *No One Was Turned Away: The Role of Public Hospitals in New York since 1900* (New York: Oxford University Press, 1999); David Rosner, *A Once Charitable Institution: Hospitals and Health Care in Brooklyn and New York, 1885–1915* (Cambridge: Cambridge University Press, 1982); Philip Shoemaker and Mary Van Hulle Jones, "From Infirmaries to Intensive Care: Hospitals in Wisconsin," in *Wisconsin Medicine: Historical Perspectives,* ed. Ronald L. Numbers and Judith Walzer Leavitt (Madison: University of Wisconsin Press, 1981), 105–31; and Morris J. Vogel, *The Invention of the Modern Hospital: Boston, 1870–1930* (Chicago: University of Chicago Press, 1980). On hospitals for African Americans, see Vanessa Worthington Gamble, *Making a Plan for Ourselves: The Black Hospital Movement, 1920–1948* (New York: Oxford University Press, 1995). On maternity and women's hospitals, see Judith Walzer Leavitt, *Brought to Bed: Child Bearing in America, 1750–1950* (New York: Oxford University Press, 1986), and Regina Morantz-Sanchez, *Conduct Unbecoming a Woman: Medicine on Trial in Turn-of-the-Century Brooklyn* (New York: Oxford University Press, 1999). David J. Rothman, *The Discovery of the Asylum* (Boston: Little, Brown, 1971) examines the early development of institutions for the care and treatment of the insane.

4. U.S. Department of Commerce, Bureau of the Census, *Benevolent Institutions 1904* (Washington DC: Government Printing Office, 1905), 184–92; Carlan Craman, O.S.F., "Women Religious in Health Care: The Early Years," in *Pioneer Healers: The History of American Women Religious in Health Care,* ed. Ursula Stepsis, C.S.A., and Dolores Liptak, R.S.M. (New York: Crossroad, 1989), 23.

5. "Who Shall Take Care of Our Sick?" *Catholic World* 8 (October 1868): 55.

6. Rev. J. R. Bayley to the Board of Governors of the New York Hospital, 6 January 1851, AANY; New York Hospital Board of Governors, *Minutes of the Monthly Meeting,* 7 January 1851, NYH.

7. *Freeman's Journal,* 11 March 1848.

8. A nun takes what are referred to as solemn vows and lives a cloistered life within a convent. For the most part, hospital sisters belonged to what are called active communities, ones where sisters engage in work outside their convents. Women religious is a contemporary term. Evangeline Thomas, ed., *Women Religious History Sources* (New York: R. R. Bowker, 1983), xxvi.

9. Rev. Walter Elliot, C.S.P., "Saint Vincent De Paul and the Sisters of Charity," *Catholic World* 70 (October 1899): 25.

10. New York City Department of Finance, *Private Charitable Institutions Receiving Public Money* (New York: Martin Brown Press, 1904), 15–32; E. H. Lewinski-Corwin, *The Hospital Situation in Greater New York* (New York: G. P. Putnam's Sons, 1924), 35–36.

11. St. Patrick's Cathedral may be the most familiar Roman Catholic image in the United States; it is the cover illustration for historian Jay P. Dolan's most recent book, *In Search of American Catholicism* (New York: Oxford University Press, 2002).

12. Sisters of Charity, *Lives,* 100–104, SCMSV; Sister Marie de Lourdes Walsh, *The Sisters of Charity of New York,* 2 vols. (New York: Fordham University Press, 1960), 1:3. For biographical information on John Hughes, see Richard Shaw, *Dagger John: The Unquiet Life and Times of John Hughes* (New York: Paulist Press, 1977); Rev. Msgr. Florence J. Cohalan, *A Popular History of the Archdiocese of New York* (Yonkers: New York Catholic Historical Society, 1983), 50–51; John Tracy Ellis, "John Hughes, Leader of the Church at Mid-Century," *Perspectives in American Catholicism* (Baltimore: Helicon, 1963), 100–107; and Charles R. Morris, *American Catholic* (New York: Vintage Books, 1997), 3–5.

## CHAPTER ONE. "A Climate New to Them"

1. *Freeman's Journal,* 4 November 1848; Carlan Craman, O.S.F., "Women Religious in Health Care: The Early Years," in *Pioneer Healers: The History of American Women Religious in Health Care,* ed. Ursula Stepsis, C.S.A., and Dolores Liptak, R.S.M. (New York: Crossroad, 1989), 23. See the appendixes in *Pioneer Healers* for a listing of names and dates relevant to the story of women religious in health care in the United States and George C. Stewart, *Marvels of Charity: History of American Sisters and Nuns* (Huntington, IN: Our Sunday Visitor, 1994), 516–48, for a listing of sister-founded hospitals.

2. Page Cooper, *The Bellevue Story* (New York: Thomas Crowell and Company, 1948), 38–39. Other histories of New York's public hospitals include Robert J. Carlisle, M.D., ed., *An Account of Bellevue Hospital* (New York: Society of the Alumni of Bellevue Hospital, 1896), BHA; Frederick M. Dearborn, *The Metropolitan Hospital* (New York, 1937); selections in Henry J. Cammann and Hugh N. Camp, *The Charities of New York, Brooklyn, and Staten Island* (New York: Hurd and Houghton, 1868); Sandra Opdycke, *No One Was Turned Away: The Role of Public Hospitals in New York City since 1900* (New York: Oxford University Press, 1999); and John Fletcher Richmond, *New York and Its Institutions* (New York: E. B. Treat, 1871). On Brooklyn's public institutions in the nineteenth century, see New York State Charities Board, "Report upon the Public Charities of King's County," *Annual Report 1880*.

3. William T. White, ed., *Medical Register of New York, New Jersey and Connecticut,* vol. 23 (New York: G. P. Putnam's Sons, 1885), 71–72; Robert Ernst, *Immigrant Life in New York* (New York: Kings Crown Press, 1949), 29.

4. Samuel Osgood, *New York in the Nineteenth Century* (New York: New-York Historical Society, 1866), 75–89; Ernst, *Immigrant Life in New York,* 53.

5. Ray Allen Billington, *The Protestant Crusade, 1800–1860* (1938; reprint, Chicago: Quadrangle Books, 1964), 35.

6. The New York Association for Improving the Condition of the Poor, *Thirteenth Annual Report for the Year 1856,* 25–26.

7. Raymond A. Mohl, *Poverty in New York, 1783–1825* (New York: Oxford University Press, 1971), 193.

8. As quoted in Charles E. Rosenberg, *The Care of Strangers* (New York: Basic Books, 1987), 46.

9. New York City Almshouse Commissioner, *Annual Report, 1856,* NYCMA.

10. Rt. Rev. John Dubois, *Pastoral Letter to the Clergy and Laity of the Diocese* (New York: John Doyle, 1834), 3, AANY.

11. *Freeman's Journal,* 11 March 1848; Francis X. Curran, S.J., *The Return of the Jesuits: Chapters in the History of the Society of Jesus in the Nineteenth Century* (Chicago: Loyola University, 1966), 104–5, SJNY.

12. New York City Almshouse Commissioner, *Annual Report 1848,* 189, NYCMA.

13. *Freeman's Journal,* 29 January 1848.

14. New York City Board of Alderman, *Proceedings,* vol. 35, pt. 1 (9 May–2 August 1848), 371, NYCMA.

15. Rev. Jeremiah W. Cummings, D.D., to Archbishop John Hughes, 29 March 1858, file A5–A11, AANY.

16. Thomas Campbell, S.J., to Msgr. T. S. Preston, Vicar General, n.d. (ca. 1890), Grant correspondence 88-GHJ-5, NYCMA; *The Woodstock Letters: A Record of the Current Events and Historical Notes Connected with the Colleges and Missions of the Society of Jesus in North and South America,* vols. 1–80 (Woodstock, MD: Woodstock College, 1872–1969), 24:401, SJNY.

17. *Freeman's Journal,* 11 March 1848.

18. State Charities Aid Association, Bellevue Hospital Visiting Committee, Committee on the Surgical Wards for Women, "Transcript of Hospital Visits," February 1872, 1; *Report of the Committee on the Surgical Wards for Women,* 5 March 1873, BHA.

19. New York Almshouse Commissioner, *Annual Report 1848,* 189, NYCMA.

20. Rev. J. R. Bayley, Secretary, to the Archbishop of New York to the Board of Governors of the New York Hospital, 6 January 1851, AANY; New York Hospital Board of Governors, *Minutes of the Monthly Meeting,* 7 January 1851, NYHA; *Freeman's Journal,* 3 September 1853. New York became an archdiocese in 1850, Brooklyn a separate diocese in 1853. Rev. Msgr. Florence D. Cohalan, *A Popular History of the Archdiocese of New York* (Yonkers: U.S. Catholic Historical Society, 1983), 71–72.

21. Rev. J. R. Bayley to the Board of Governors of the New York Hospital, 6 January 1851, NYHA.

22. For a discussion of how Irish and German immigrants viewed these sacraments, see Jay P. Dolan, *The Immigrant Church: New York's Irish and German Catholics, 1815–1865* (Notre Dame: University of Notre Dame Press, 1983), chaps. 3 and 4.

23. Nelson J. Callahan, *The Diary of Richard L. Burtsell* (New York: Arno, 1978), 27.

24. Robert Emmett Curran S.J., *Michael Augustine Corrigan and the Shaping of Conservative Catholicism in America, 1878–1902* (New York: Arno, 1977), 176.

25. Callahan, *Diary of Richard Burtsell,* 55. The need for a local institution to train priests was a goal of the hierarchy that was finally achieved in 1896 with the opening of Saint Joseph's Seminary in Yonkers; Cohalan, *Archdiocese of New York,* 163–64.

26. Dolan, *Immigrant Church,* 57.

27. Ibid., 58.

28. Callahan, *Diary of Richard Burtsell,* 151–52.

29. *Woodstock Letters,* 3:174.

30. Ibid., 23:368–69.

31. Ibid., 23:92.

32. Ibid., 3:174.

33. *Metropolitan Record,* 29 June 1861.

34. *Metropolitan Record,* 14 March 1863; 21 March 1863.

35. *Woodstock Letters,* 1:59.

36. Ibid., 23:84, 24:401; Rev. John K. Sharp, *History of the Diocese of Brooklyn, 1853–1953,* 2 vols. (New York: Fordham University Press, 1954), 1:210; *McKinney's Consolidated Laws of New York* (St. Paul: West Publishing, 1987), 402.

37. Rev. E. J. Crawford, *The Daughters of Dominic on Long Island,* 2 vols. (New York: Benziger Brothers, 1938, 1953), 1:250–53; Cohalan, *Archdiocese of New York,* 61, 158; Dolan, *Immigrant Church,* 131–32; Pietro Di Donato, *Immigrant Saint* (New York: McGraw-Hill, 1960), 110.

38. Ethnic parishes were characteristic of Catholicism throughout the United States in the nineteenth century. For a general discussion, see Dolan, *The American Catholic Experience: A History from Colonial Times to the Present* (Garden City, NY: Doubleday, 1985), chap. 6. For the history of specific immigrant parishes in New York City, see Mary Elizabeth Brown, "Italian Immigrants and the Catholic Church in the Archdiocese of New York, 1880–1950" (Ph.D. diss., Columbia University, 1987), 5–64, 155–210, and Dolan, *Immigrant Church,* 45–87, which discusses Irish and German parishes.

39. Brown, "Italian Catholics in the Archdiocese of New York," 50–54, 148–53; Crawford, *Daughters of Dominic,* 1:18–24; Dolan, *Immigrant Church,* 13. Within their own churches, groups continued specific religious traditions. For example, while Mary was an important figure to all Catholics, churches of varying ethnicities maintained very different ways of expressing their devotion to her. The German parishes continued special customs and prayers that had originated in Germany in the eighteenth century; ceremonies, prayers, and even statues were brought from churches back home. Later, Italian Catholics expressed their devotion to Mary in celebrations brought with them as well.

Dolan, *Immigrant Church,* 76; Robert Orsi, *The Madonna of 115th Street: Faith and Community in Italian Harlem, 1880–1950* (New Haven: Yale University Press, 1985), 51–52. On nineteenth-century American Catholic devotions, see Ann Taves, *The Household of Faith: Roman Catholic Devotions in Mid-Nineteenth-Century America* (Notre Dame: University of Notre Dame Press, 1986).

40. Dolan, *Immigrant Church,* 69.

41. Brown, "Italian Catholics in the Archdiocese of New York," 110.

42. Missionary Sisters of the Sacred Heart, Opere Della Madre Cabrini, May 1910, 4, translation, MSSH.

43. Claims of the Fathers and the Congregation of the Church of the Most Holy Redeemer, translation, file A-29, AANY.

44. "Hospital Life in New York," *Harper's New Monthly Magazine* 57 (July 1878): 182.

45. State Charities Aid Association, *Report of the Committee Appointed to Take Active Measures in Regard to the Erection of a New Bellevue Hospital* (American Church Press Company, 1874), 15.

46. New York State Charities Board, *Annual Report 1880,* 202.

47. Dolan, *Immigrant Church,* 75.

48. *Freeman's Journal,* 14 September 1850; Samuel Ward Francis, *Biographical Sketches of Distinguished Living New York Surgeons* (New York: J. Bradburn, 1866), 42–43, 109, 177–78; Sister Marie de Lourdes Walsh, *With a Great Heart: The Story of St. Vincent's Hospital and Medical Center of New York* (New York: Saint Vincent's Hospital and Medical Center of New York, 1965), 12–17.

49. *Metropolitan Record,* 12 April 1862; 19 April 1862.

50. *Metropolitan Record,* 12 April 1862.

51. *Metropolitan Record,* 19 April 1862.

52. James W. Fraser, *Between Church and State: Religion and Public Education in a Multicultural America* (New York: St. Martin's, 1999), 52–57; Vincent P. Lannie, *Public Money and Parochial Education: Bishop Hughes, Governor Seward, and the New York School Controversy* (Cleveland: Press of Case Western University, 1968), 246–47.

53. *Freeman's Journal,* 26 January 1850.

54. *Freeman's Journal,* 19 April 1856.

55. *Ibid.*

56. *Freeman's Journal,* 12 June 1858.

57. U.S. Department of Commerce, Bureau of the Census, *Benevolent Institutions 1904* (Washington, DC: U.S. Government Printing Office, 1905), 180–84.

## CHAPTER TWO. "To Serve Both God and Man"

1. *Metropolitan Record,* 12 March 1864.

2. Published data on the founding dates of New York hospitals include E. H. Lewinski-Corwin, *The Hospital Situation in Greater New York* (G. P. Put-

nam's Sons, 1924); Edwin M. Grout, "Private Charitable Institutions Receiving Public Money in New York City" (New York City: Martin Brown Press, 1904); and John F. Richmond, *New York and Its Institutions* (New York and E. B. Treat, 1872).

3. "Who Shall Take Care of Our Sick?" *Catholic World* 8 (October 1866): 42. On healing and health care among early Christians and in the Middle Ages, see Gary B. Ferngren, "Early Christianity as a Religion of Healing," *Bulletin of the History of Medicine* 66 (Spring 1992): 1–15; Darrel W. Amundsen, "The Medieval Catholic Tradition," and Marvin R. O'Connell, "The Roman Catholic Tradition since 1545," in *Caring and Curing: Health and Medicine in the Western Religious Tradition,* ed. Ronald Numbers and Darrel W. Amundsun (Baltimore: Johns Hopkins University Press, 1988), 65–145; Christopher J. Kauffman, *Ministry and Meaning: A Religious History of Health Care in the United States* (New York: Crossroad, 1995), 11–26; Guenter Risse, *Mending Bodies, Saving Souls: A History of Hospitals* (New York: Oxford University Press, 1999), 69–165.

4. For details concerning the work of the first Catholic sisterhoods in the United States, see Elinor Tong Dehey, *Religious Orders of Women in the United States* (Hammond, IN: W. B. Conkey, 1930); Barbara Misner, "A Comparative Study of the Members and Apostolates of the First Eight Permanent Communities of Women Religious within the Original Boundaries of the United States" (Ph.D. diss., Catholic University, 1980); and Sister Mary Christina Sullivan, "Some Non-Permanent Foundations of Religious Orders and Congregations," *U.S. Catholic Historical Society Records and Studies* 31 (1940): 7–118.

On nursing and early hospital work by Catholic sisterhoods in the United States, see Ann Doyle, "Nursing by Religious Orders in the United States," *American Journal of Nursing* 29 (July 1929): 775–87, (August 1929): 959–69, (September 1929): 1085–95; Carlan Kraman, "Women Religious in Health Care: The Early Years," in *Pioneer Healers: The History of American Women Religious in Health Care,* ed. Ursula Stepsis and Dolores Liptak (New York: Crossroad, 1989), 1–39; and John O'Grady, *Catholic Charities in the United States* (National Conference of Catholic Charities, 1930).

On the foundation of the Sisters of Charity in the United States, see Sister Marie de Lourdes Walsh, *The Sisters of Charity of New York,* 3 vols. (New York: Fordham University Press, 1960), 1:6–8, and William Jarvis, "Mother Seton's Sisters of Charity" (Ph.D. diss., Columbia University, 1984), 71–103.

5. Misner, "First Eight Permanent Foundations of Women Religious," 31–32, 212, 250.

6. On the history of the early American Catholic Church, see Jay P. Dolan, *The American Catholic Experience: A History from Colonial Times to the Present* (Garden City, NY: Doubleday, 1985), 101–23; Thomas T. McAvoy, "The Catholic Minority in the United States, 1789–1821," in *Early American Catholicism, 1643–1820,* ed. Timothy Walch (New York: Garland, 1988), 255–72; and John Gilmary Shea, *A History of the Catholic Church in the United States,* 4 vols. (New York: John G. Shea, 1886–1892), 1, 2.

On the early history of the Catholic Church in New York City, see "Early History of the Catholic Church on the Island of New York," *Catholic World* 10 (1869): 413–20; Shea, *Catholic Church in the United States,* 3:160–206, 495–543; and John Talbot Smith, *The Catholic Church in New York,* 2 vols. (New York: Hall and Locke, 1908). John Gilmary Shea, ed., *The Catholic Churches of New York City* (New York: L. G. Goulding and Co., 1878) is an illustrated collection of the history of individual parishes in New York City.

7. Contemplative communities are those where the sisters' lives "are organized to develop the work of prayer"; Evangeline Thomas, ed., *Women Religious History Sources: A Guide to Repositories in the United States* (New York: R. R. Bowker, 1983), xxv.

8. For a discussion of some of the earliest efforts to organize a new kind of religious community for women, see Jo Ann Kay McNamara, *Sisters in Arms: Catholic Nuns through Two Millennia* (Cambridge, MA: Harvard University Press, 1996), 452–72. On the development of modern religious communities in France, see Sarah A. Curtis, *Educating the Faithful: Religion, Schooling, and Society in Nineteenth-Century France* (DeKalb: Northern Illinois University Press, 2002). Studies of the emergence of active Irish sisterhoods include Catriona Clear, *Nuns in Nineteenth-Century Ireland* (Dublin: Gill and Macmillan, 1988); Mary Peckham Magray, *The Transforming Power of the Nuns: Women, Religion, and Cultural Change in Ireland, 1750–1900* (New York: Oxford University Press, 1998); and Tony Fahey, "Nuns in the Catholic Church in Ireland in the Nineteenth Century," in *Girls Don't Do Honors: Irish Women in Education in the Nineteenth and Twentieth Centuries,* ed. Mary Cullen (Dublin: Argus Press, 1987): 7–30.

9. On the general history of women religious in the United States, see Mary Ewens, O.P., *The Role of the Nun in Nineteenth-Century America* (New York: Arno, 1978), and "Women in the Convent," in *American Catholic Women: A Historical Exploration,* ed. Karen Kennelly, C.S.J. (New York: Macmillan, 1989), 17–47; George C. Stewart, *Marvels of Charity: History of American Sisters and Nuns* (Huntington, IN: Our Sunday Visitor, 1994); and Margaret Susan Thompson, "Discovering Foremothers: Sisters, Society, and the American Catholic Experience," *U.S. Catholic Historian* 5 (Summer/Fall 1986): 273–91.

On Roman Catholic and Protestant nursing orders in the United States, see Doyle, "Nursing by Religious Orders," 775–86, 959–69, 1085–95, 1331–43, 1467–84; Stepsis and Liptak, *Pioneer Healers*; Sister Mary Denis Maher, *To Bind Up the Wounds: Catholic Sister Nurses in the Civil War* (New York: Greenwood, 1989). Sioban Nelson, *Say Little, Do Much: Nursing, Nuns and Hospitals in the Nineteenth Century* (Philadelphia: University of Pennsylvania Press, 2001), is a comparative study of Roman Catholic nursing sisters in Australia, England, and the United States.

General histories of the New York congregations involved in hospital work include Rev. E. J. Crawford, *The Daughters of Dominic on Long Island,* 2 vols. (New York: Benziger Brothers, 1938 and 1953); Sister M. Pauline Hill, *In Love with Christ's Poor* (Cincinnati: St. Clare's Provincial House, 1959); Sister Mary Ignatius Meany, C.S.J., *By Railway or Rainbow: A History of the Sisters of Saint Joseph Brentwood* (Brentwood, NY: Pine Press, 1964); and Walsh, *Sisters of Charity.*

10. Sisters of Charity Mount St. Vincent, *Constitution of the Sisters of Charity of Saint Vincent de Paul of New York* (New York: n.d.), 7, SCMSV; Ewens, *Role of the Nun,* 127–28; Walsh, *Sisters of Charity,* 1:127–45.

11. The Brooklyn Dominicans were initially organized in Germany as a congregation of the Second Order, which under canon law was a cloistered group. In 1896, the American sisters changed their congregation's status to acknowledge their noncloistered work in teaching and nursing. Crawford, *Daughters of Dominic,* 1:307–9.

12. Thomas, *Women Religious History Sources,* xxvii.

13. Patricia Byrne, C.S.J., "Sisters of St. Joseph: The Americanization of a French Tradition," *U.S. Catholic Historian* 5 (Summer/Fall 1986): 242–49. On religious women in France in the seventeenth century, see Elizabeth Rapley, *The Devotes* (Montreal: McGill-Queens University Press, 1980).

14. Sisters of Charity Mount St. Vincent's, *Regulations for the Society of the Sisters of Charity in the United States of America* (New York: n.d. [ca.1880]), n.p., SCMSV.

On the origins of the Daughters of Charity and their nursing and hospital work in France in the seventeenth and eighteenth centuries, see Colin Jones, *The Charitable Imperative* (London: Routledge, 1989.)

15. Sisters of Charity Mount St. Vincent's, *Regulations.*

16. Thompson, "Discovering Foremothers," 287.

17. "Statutes for the Sisters of the Poor of St. Francis in Aachen," no. 27, in Letters of Mother Frances Shervier, copied from the archives of the congregation in Aachen for St. Clare's Province, Hartwell, Cincinnati, trans. Sister Pauline Hill, n.d., record group 2, FSP.

18. Francis Shervier to Sister Felicitas, 1862, in Shervier, Letters, 202; "A Chronology of the Franciscan Sisters of the Poor, 1858–1983," ts., 1983; "Activities of a Century in the United States," ts., 1958, n.p., FSP.

19. Sisters of Charity Mount St. Vincent's, *Regulations.*

20. See Misner, "First Eight Permanent Communities of Women Religious," and Sister Mary Christine Sullivan, "Some Non-Permanent Foundations of Religious Orders and Congregations," *U.S. Catholic Historical Society Records and Studies* 31 (1940): 7–18, for further discussion of the earliest religious communities in the United States.

21. Ewens, *Role of the Nun,* 102–3, "Women in the Convent," 17; Charles E. Rosenberg, *The Cholera Years: The United States in 1932, 1849, and 1866* (Chicago: University of Chicago Press, 1962), 64, 95, 139–40.

22. Ray Allen Billington, *The Protestant Crusade, 1800–1860* (1938; reprint, Chicago: Quadrangle Books, 1964), 413–15; James R. Lewis, "'Mind-Forged Manacles': Anti-Catholic Convent Narratives in the Context of the American

Captivity Tale Tradition," *Mid-America* 72 (October 1990): 157–60; Joseph G. Mannard, "Maternity . . . of the Spirit: Nuns and Domesticity in Antebellum America," *U.S. Catholic Historian* 5 (Summer/Fall 1986): 306–9; Nancy Lusignan Schultz, *Fire and Roses: The Burning of the Charlestown Convent, 1834* (New York: Free Press, 2000); Barbara Welter, "From Maria Monk to Paul Blanshard," in *Uncivil Religion: Interreligious Hospitality in America,* ed. Robert N. Bellah and Frederick Greenspahn (New York: Crossroad, 1987), 47–51.

23. Mannard, "Maternity of the Spirit," 308–11; Amanda Porterfield, *Female Spirituality* (Philadelphia: Temple University Press, 1980), 16–17; Welter, "Maria Monk," 53–57.

24. As quoted in Ewens, *Role of the Nun,* 224. Both northern and southern Catholics were concerned about the religious fate of Catholic soldiers in their armies. The New York Catholic press complained about the lack of Catholic chaplains in Union Army hospitals and the Sisters of Charity, who worked with Confederate officers, complained that they were watched very closely "for fear we might elude from their group, one of those poor souls and bring them into the true fold." Another sister was disturbed about the poor religious state of most of the soldiers. Like New York City's immigrants, many were only nominally Catholic "for many had forgotten their duties as such, but it was our consolation to see them entering upon them again with the simplicity of children." Presumably the battlefield hospital, like others, offered opportunities for conversion. In this case, "about 50 were baptized." John Tracy Ellis, ed., *Documents of American Catholic History* (Wilmington, DE: Michael Glazier, 1987), 1:378; *Metropolitan Record,* 2 July 1864; "Notes on the Hospital of Warrington, Florida," *Pensacola Historical Society Quarterly* 3 (July 1967): n.p.

25. John R. G. Hassard, *Life of the Most Reverend John Hughes, First Archbishop of New York, with Extracts from His Private Correspondence* (New York: D. Appleton and Co., 1866), 441–42, AANY.

26. "Notes on the Hospital of Warrington, Florida"; Ellis, *Documents,* 376–78; Ewens, *Role of the Nun,* 222–24; Herron, *Sisters of Mercy,* 34–39; Jolly, *Nuns of the Battlefield,* 208–19; Maher, *To Bind Up the Wounds,* 69; Mary Elizabeth Massey, *Bonnet Brigades* (New York: Alfred A. Knopf, 1966), 47.

27. *Metropolitan Record,* 23 August 1862; Herron, *Sisters of Mercy,* 34–39; Jolly, *Nuns of the Battlefield,* 35, 208–12; Maher, *To Bind Up the Wounds,* 105–7.

28. Roy Rosenzweig and Elizabeth Blackmar, *The Park and the People: A History of Central Park* (Ithaca: Cornell University Press, 1992), 75; Page Cooper, *The Bellevue Story* (New York: Thomas Y. Crowell and Company, 1948), 78; Jolly, *Nuns of the Battlefield,* 25–6; Maher, *To Bind Up the Wounds,* 74; Walsh, *Sisters of Charity,* 1:191, 3:170, 175.

29. Joseph R. Smith by order of the Surgeon General to Charles McDougall, 11 September 1862, and W. A. Hammond Surgeon General to Mr. Van Buren, 17 October 1862, Letterbook, SCMSV; Walsh, *Sisters of Charity,* 3:166–75.

30. Joseph R. Smith to Charles A. McDougell, 10 September 1862, Letterbook, SCMSV.

31. David L. Gollaher, *Voice for the Mad* (New York: Free Press, 1995), 413–14.

32. Cecil Woodham-Smith, *Florence Nightingale* (Edinburgh: Constable, 1950), 103, 108, 114, 143–44, 184.

33. Gollaher, *Voice for the Mad,* 413–14.

34. John Hill Brinton, *Personal Memoirs of John Hill Brinton* (New York: Neale, 1914), 44; Maher, *To Bind Up the Wounds,* 128–31.

35. Ewens, *Role of the Nun,* 221.

36. Crawford, *Daughters of Dominic,* 1:262; Walsh, *Sisters of Charity,* 1:206–7.

37. Charles Renaud,"Proposition au Conseil d'Administration de la Societe Francaise de Bienfaisance de New York," 4 December 1885, French Hospital collection, NYAM. The Congregation of the Marianites of the Holy Cross was founded in Le Mans, France, in 1841. The first sisters came to the United States, to Notre Dame, Indiana, in 1843. Thomas, *Women Religious History Sources,* 50. In hiring sisters, the French immigrants were following French tradition. In France, sisters remained in hospitals in the nineteenth century despite anticlerical opposition. See Katrin Schultheiss, *Bodies and Souls: Politics and the Professionalization of Nursing in France, 1880–1922* (Cambridge, MA: Harvard University Press, 2001), 3–4.

38. *Freeman's Journal,* 3 December 1853.

39. "Who Shall Take Care of Our Sick?" *Catholic World* 8 (October 1868): 46.

40. Franciscan Sisters of the Poor, "History of St. Francis' Hospital in New York, 1865–18 (and to 1894)," ts., 1936, 1:4–5, FSP; Mary Immaculate Hospital Receipts and Expenditures Book 1902–13, Mary Immaculate Hospital file, SSDCHC; "Who Shall Take Care of Our Sick?" 45.

41. Notes of Managers Meeting, 13 October 1916, and letter to Sister Mary Ignatius, 12 September 1918, St. Lawrence Hospital Letterbook, vol. 2, SCMSV; Sister Epiphania, Superintendent St. Francis' Hospital, to Cardinal Farley, 6 February 1907, St. Francis' Hospital file, AANY; Walsh, *Sisters of Charity,* 3:133.

42. St. Vincent's Hospital, *Thirty-Second Annual Report for the Year Ending September 30, 1881,* 522v-2a, SCMSV. Sisters of St. Dominic, Lives of Deceased Sisters, 17 vols., Lives of the Mothers, 2 vols. (Brentwood, NY: n.d.), SSDCHC; Sisters of Charity Mount St. Vincent's, Lives of Our Departed Mothers, 103, SCMSV.

43. Misericordia Sisters, *Historical Survey of the Mission of the Misericordia Sisters* (Montreal: Misericordia Sisters, 1980), n.p.; Rev. Msgr. Florence D. Cohalan, *A Popular History of the Archdiocese of New York* (Yonkers: U.S. Catholic Historical Society, 1983), 93, 158; Crawford, *Daughters of Dominic,* 1:3; James P. Dolan, *The Immigrant Church: New York's Irish and German Catholics, 1815–1860* (Notre Dame: University of Notre Dame Press, 1983), 131.

44. Crawford, *Daughters of Dominic,* 1:17–50.

45. *Truth Teller,* 5 March 1831; Ewens, *Role of the Nun,* 67, 134–35; Walsh, *Sisters of Charity,* 3:20.

46. Sisters of Charity Mount St. Vincent's, Lives, 100–104, SCMSV; Walsh, *Sisters of Charity,* 3:133.

47. "List of Sisters Who Have Served at St. Vincent's Hospital," ts., 1977, 522v-2d, SCMSV; Franciscan Sisters of the Poor, The Chronicles of St. Peter's Hospital Brooklyn, New York, vol. 1, introduction, Record Group 6, Series: St. Peter's Hospital, Religious Personal Record Group no. 10, FSP; Sisters of St. Dominic, Lives of Sisters, 1:63–64, 2:232–33, 15:106–8; Crawford, *Daughters of Dominic,* 1:136, 2:appendix; Walsh, *Sisters of Charity,* 3:133.

48. *Metropolitan Record,* 7 May 1859; 14 May 1859.

49. Sisters of Charity, Lives, 85.

50. Sr. Mary Loretto Donahue, personal file, SCMSV; Sisters of St. Dominic, Lives, 4:63–64; Jay P. Dolan, *The Immigrant Church,* 106; John K. Sharp, *Priests and Parishes of the Diocese of Brooklyn* (New York: Loughlin Brothers, 1944), 111.

51. Sisters of Charity, Lives, 5–9; "List of St. Vincent's Hospital and Medical Center of Manhattan Administrators, 1977," ts., ca. 1977, 522v-2e: SCMSV.

52. Thomas P. McCarthy, C.S.V., *Guide to the Catholic Sisterhoods in the United States* (Washington: Catholic University of America, 1964), 390–91; Thomas, *Women Religious History Sources,* xxv–xxvii.

53. Mother Seraphine Staimer, "Founding of the Convent of the Dominican Sisters (1853) at the Holy Trinity Church in Williamsburg with Its Dependent Convents," vol. 1 [ca. 1865], 13, SSDCHC; Sisters of St. Dominic, Lives, 1:20–21, 2:247–48, 445–46.

54. Sisters of St. Dominic, Lives, 6:1430–38; Sisters of Charity, Lives, 85; Walsh, *Sisters of Charity,* 1:183.

55. Staimer, "Dominican Sisters," 1:34.

56. Ibid., 1:36–37.

57. Sisters of St. Dominic, Lives of Mothers, 1; *Hospital Progress* 12 (May 1931): 230.

58. Sister M. Jerome to Halifax, Palm Sunday, 1857, Letterbook, SCMSV; Sister Mercedes Hess to Sister Polycarp Larsson, 25 December 1907, St. Catherine's Hospital file, SSDCHC; Shervier, Letters, 55–56.

59. Mother Frances Shervier to Sisters at home from Cincinnati, 21 July 1863, Shervier, Letters.

60. Ibid.

61. "Gleanings from the Life of Sister Mary Reginald Kerling O.P. 1853–1936 by a Sister of St. Dominic," ts., n.d., St. Catherine's Hospital file, SSDCHC.

62. "Origins and Customs and Observations in Our Congregation Collected from Our Annals," in Shervier, Letters, 55–56.

63. Mother Frances to Sister Felicitas, 1859, Shervier, Letters, 199.

64. Mother Frances Schervier to Sister Augustin Keussen, 1863, Shervier, Letters, 75.

65. Ibid.; "Gleanings from the Life of Sister Mary Reginald Kerling," 11–12, SSDCHC; Crawford, *Daughters of Dominic,* 1:301–8.

66. Sisters at St. John's Hospital, photograph [ca. 1900], St. John's Hospital file, ASSJ; Patricia Byrne, C.S.J., "Sisters of St. Joseph: The Americanization of a French Tradition," *U.S. Catholic Historian* 5 (Summer/Fall 1986): 268–70.

67. Susan Reverby, *Ordered to Care: The Dilemma of American Nursing* (Cambridge: Cambridge University Press, 1987), 12, 22.

68. Ibid., 39–61.

69. State Charities Aid Association, *A Century of Nursing with Hints toward the Organization of a Nurses Training School* (New York: G. P. Putnam Sons, 1876), 36.

70. Dorothy Giles, *A Candle in Her Hand* (New York: G. P. Putnam's Sons, 1949), 162–63.

71. Sister Magda Marie to Rev. John Hunt, 21 July 1971, Sister M. Rose Bernadette Connolly, P.P., "The Sisters of St. Dominic, Brooklyn, New York, Their Interest in Nursing Education" (B.S. thesis, St. Louis University, 1940), St. Catherine's Hospital file, SSDCHC; Sister Marie Le Gras Byrne, *A History of St. Vincent's Hospital School of Nursing* (Washington, DC: Catholic University Press, 1941), 11–12. Founding dates cited for nursing schools can vary. Some citations refer to starting dates of programs that were short-lived, others permanent founding dates, others the date a school received state certification. St. Mary's is cited as 1889 in Ann Doyle, "Nursing by Religious Orders in the United States," *American Journal of Nursing* 29 (September 1929): 1085–95.

72. Renaud, "Proposition."

73. Ibid.

## CHAPTER THREE. "Consoling Influences"

1. St. John's Hospital, *Record Book 1891–1895,* St. John's Hospital file, SSJB.

2. On the role of the nineteenth-century hospital in medical treatment, see Charles E. Rosenberg, *The Care of Strangers: The Rise of America's Hospital System* (New York: Basic Books, 1987) and "And Heal the Sick: The Hospital and the Patient in 19th-Century America," *Journal of Social History* (June 1977): 428–47.

3. On therapeutic practices in this period, see Charles E. Rosenberg, "The Practice of Medicine in New York a Century Ago" in *Sickness and Health in America: Readings in the History of Medicine and Public Health,* ed. Judith Walzer Leavitt and Ronald L. Numbers (Madison: University of Wisconsin Press, 1978), 55–74, and "The Therapeutic Revolution: Medicine, Meaning, and Social Change in Nineteenth-Century America," in *The Therapeutic Revolution,* ed. Morris Vogel and Charles E. Rosenberg (Philadelphia: University of Pennsylvania Press, 1979), 3–25; and John Harley Warner, *The Therapeutic Perspective: Medical Practice, Knowledge and Identity in America, 1820–1885* (Cambridge, MA: Harvard University Press, 1986).

4. Jewish hospitals made up the next highest total among religious hospitals: there were twelve Jewish hospitals with a total of 1,800 beds. New York City Department of Finance, *Private Charitable Institutions Receiving Public Money in New York* (New York: Martin Brown Press, 1904), 15–32; E. H. Lewinski-Corwin, *The Hospital Situation in Greater New York* (New York: G. P. Putnam's Sons, 1924), 21–25, 36; U.S. Department of Commerce, Bureau of the Census, *Benevolent Institutions 1904* (Washington, DC: U.S. Government Printing Office, 1905), 180–84.

To see this increase in the number of Catholic hospitals in a national perspective, see Rosenberg, *Care of Strangers,* 109–10. On the development of Catholic hospitals in Philadelphia in this period, see Gail Carr Fasterline, "Saint Joseph's and Saint Mary's: The Origins of Catholic Hospitals in Philadelphia," *The Pennsylvania Magazine of History and Biography* 108 (July 1984): 289–314. Tina Levitan, "Rise of the Jewish Hospitals of New York City," *The New York State Journal of Medicine* 15 (December 1964): 3027–32, surveys the origins of several Jewish hospitals in New York.

5. St. Vincent's Hospital of New York, *Fourteenth Annual Report for the Year 1863,* file 522v-2a, SCMSV. "Who Shall Take Care of Our Sick?" *Catholic World* 8 (October 1868): 46; Henry J. Cammann and H. N. Camp, *The Charities of New York, Brooklyn, and Staten Island* (New York: Hourd and Houghlin, 1868), 47.

6. *Metropolitan Record,* 30 April 1859.

7. St. Vincent's Hospital of New York, *Annual Report 1863.* Studies of the treatment of tuberculosis in the nineteenth century include the classic by Rene Dubos and Jean Dubos, *The White Plague: Tuberculosis, Man, and Society* (1952; reprint, New Brunswick: Rutgers University Press, 1987); and more recently Barbara Bates, *Bargaining for Life: A Social History of Tuberculosis, 1876–1938* (Philadelphia: University of Pennsylvania Press, 1992); and Sheila M. Rothman, *Living in the Shadow of Death: Tuberculosis and the Social Experience of Illness in American History* (New York: Basic Books, 1994).

8. Certificate of Incorporation St. Mary's Maternity and Infant's Home of the City of Brooklyn, 18 April 1888, Minutes of St. Mary's Maternity and Infant's Home, May 1893–1903, file 513M, SCMSC; New York Mothers Home of the Sisters of the Misericorde, *Annual Report 1894, Annual Report 1895,* and Misericordia Hospital, *Annual Report 1906,* Misericordia Hospital file, AANY; Our Lady of Mercy Medical Center, *A Guide to Health Services* (Bronx, NY: Our Lady of Mercy Medical Center, n.d. [ca. 1985]), n.p.: author's collection; New York State Board of Charities, *Minutes,* 7 November 1888, p. 175, 10 June 1909, p. 86; Sister Marie de Lourdes Walsh, *The Sisters of Charity of New York,* 3 vols. (New York: Fordham University Press, 1960), 3:206–7, 220.

For an analysis of some of these efforts within the context of social welfare in New York City and Irish immigrant women, see Maureen Fitzgerald, "Irish-Catholic Nuns and the development of New York City's Welfare System 1840–

1900" (Ph.D. diss., University of Wisconsin, 1992), and Mary C. Kelly, "'Forty Shades of Green': Conflict over Community among New York's Irish, 1860–1920" (Ph.D. diss., Syracuse University, 1997).

9. "Who Shall Take Care of Our Sick?" *Catholic World* 8 (October 1868): 46.

10. John Fletcher Richmond, *New York and Its Institutions* (New York: E. B. Treat, 1971), 377.

11. Sioban Nelson, *Say Little, Do Much: Nursing, Nuns, and Hospitals in the Nineteenth Century* (Philadelphia: University of Pennsylvania Press, 2001), 2–3. The first Lutheran Deaconesses came to the United States from Germany in 1849 to work at a hospital in Pittsburgh. In New York, Deaconess nurse Sister Elizabeth Fedde came from Oslo to work among the sick poor in Brooklyn in 1883, and her efforts were extended with the organization of Deaconess Hospital. Ann Doyle, R.N., "Nursing by Religious Orders in the United States," *American Journal of Nursing* 29, 10–11 (1929): 1197, 1203. The Episcopal Sisters of St. Mary opened St. Mary's Free Hospital for Children in Manhattan in 1870. Prior to that, women of a group called the Sisterhood of the Holy Communion, later formally organized as a sisterhood, were involved at St. Luke's Hospital when it opened in 1852. From there, sisters went on to St. John's Hospital in Brooklyn in 1871. Doyle, "Nursing by Religious Orders," 12:1466–71. On the development of nursing and the growth of Anglican sisterhoods in England, see Michael Hill, *The Religious Order* (London: Heinemann Educational Books, 1973).

12. Rosenberg, *Care of Strangers,* 307.

13. Rennie B. Schoepflin, *Christian Science on Trial: Religious Healing in America* (Baltimore: Johns Hopkins University Press, 2003), explains the development of Christian Science healing as an alternative medicine and the opposition it faced from medical and state authorities.

14. *Freeman's Journal,* 26 January 1850.

15. Sister M. Jeanette Jacke, O.P., "History of Mary Immaculate Hospital," ts. [ca. 1971], *New York Herald,* Brooklyn Supplement, 22 May 1904, clipping, Mary Immaculate Hospital file, SSDCHC; *Freeman's Journal,* 14 September 1850; Samuel Ward Francis, *Biographical Sketches of Distinguished Living Surgeons* (New York: J. Bradburn, 1866), 42–43, 109, 177–78; James G. Wilson, *History of New York City* (New York: n.p., 1892–93), 402; Sister Marie De Lourdes Walsh, *With a Great Heart: The Story of St. Vincent's Hospital and Medical Center of New York, 1849–1964* (New York: St. Vincent's Hospital and Medical Center of New York, 1965), 12–20.

16. *Metropolitan Record,* 23 April 1859.

17. *Metropolitan Record,* 26 April 1862.

18. *Metropolitan Record,* 12 October 1871.

19. "Private Charities and Public Money," *Catholic World* 29 (May 1879): 278.

20. Richmond, *New York and Its Institutions,* 377.

21. *Freeman's Journal,* 1 December 1849.

22. It was hoped that advertising this would be financially rewarding; the article went on to note that the policy was particularly "appreciated by the Jews of the district . . . [they] are always generous when appealed to for contributions"; "Charities of New York City," *Catholic World* 9 (September 1886): 812.

23. "Private Charities and Public Money," *Catholic World* 29 (May 1879): 278.

24. Cammann and Camp, *Charities of N.Y., Brooklyn, and S.I.,* 47.

25. Amid the usual praise among Catholics about Catholic hospitals there was at least one critic. Rev. Edward McGlynn, pastor of Saint Stephen's Roman Catholic Church on 28th Street and an active participant in both church and city politics, included Catholic hospitals in his general disapproval of all Catholic charity institutions and organizations. In 1886 the iconoclastic McGlynn explained to readers of the *New York Tribune* that "You may go forever with your hospitals and orphan asylums and Saint Vincent de Paul Societies but with them you can't cure the trouble. They relieve not eradicate. In a right state of society there ought not to be any hospitals or asylums or charitable societies." McGlynn's opinion on Catholic charity was printed in the midst of a controversy with Archbishop Michael Corrigan, which resulted in McGlynn's excommunication from 1887 to 1893 after he had enthusiastically supported the mayoral campaign of Henry George against Corrigan's instructions. Along with hospitals and orphanages, McGlynn also disapproved of Catholic schools and he was an avid Republican during the Civil War and Reconstruction. His comments in November 1886 suggest the depth of his theological disagreement with church officials. "So long as ministers of the gospel and priests of the Church tell the hard working poor to be content with their lot and hope for good times in heaven, so long will skepticism increase"; *New York Tribune,* 26 November 1886.

26. *Freeman's Journal,* 7 August 1858.

27. St. John's Hospital Long Island City, *Annual Report 1906–1907, Annual Report 1908–1909,* St. John's Hospital file, SSJB.

28. Case R7, Community Service Society Collection, CUA.

29. St. John's Hospital, Record Book 1891–95, St. John's Hospital file, ASSJB; St. Francis' Hospital, *Fortieth Annual Report for the Year Ending 1905,* RPA; "Who Shall Take Care of Our Sick?" 45.

30. "Who Shall Take Care of Our Sick?" 46.

31. New York State Board of Charities, *Directory of the Charitable Institutions of New York State* (New York: April 1892), 315.

32. St. Vincent's Hospital of the City of New York, *Thirty-Second Annual Report for the Year Ending September 30, 1881, Thirty-Ninth Annual Report for the Year 1888, Fortieth Annual Report for the Year 1889, Forty-Second Annual Report for the Year 1891, Forty-Third Annual Report, Forty-Fourth Annual Report for the Year 1894, Forty-Fifth Annual Report for the Year 1894, Forty-Sixth Annual Report for the Year 1895, Forty-Seventh Annual Report for the Year 1896, Fifty-First Annual*

*Report for the Year 1900,* file 522v-2a, SCMSV; St. Francis' Hospital, *Annual Report 1905.*

33. Rosenberg, *Care of Strangers,* 122–27, 137–41.

34. John F. Byrne, C.S.S.R.; *The Redemptorist Centenaries* (Philadelphia: Dolphin Press, 1932), 148, RPA. Walsh, *Sisters of Charity,* 3:133.

35. Metropolitan Health Board of the State of New York, *2nd Annual Report* (New York: Union Printing House, 1868); Claims of the Fathers and the Congregation of the Church of the Most Holy Redeemer, file A-29, AANY; Jay P. Dolan, *The Immigrant Church: New York's Irish and German Catholics, 1815–1965* (Notre Dame: University of Notre Dame Press, 1983), 29.

36. New York State Board of Charities, *Directory* (1892), 295–301, 313–31.

37. Rosenberg, *Care of Strangers,* 286–87.

38. Seton Hospital, *Annual Report for the Year Ending 1896,* file 519S, SCMSV; Walsh, *Sisters of Charity,* 3:213–14.

39. For a good summary of the characteristics of the American hospital in 1900, see Rosemary Stevens, *In Sickness and in Wealth: American Hospitals in the Twentieth Century* (New York: Basic Books, 1989).

40. St. John's Hospital, Record Book, 1891–1894, St. John's Hospital file, SSJB.

41. St. Vincent's Hospital of New York, *Annual Report 1881, Annual Report 1900.*

42. St. Vincent's Hospital of New York, *Annual Report 1881.*

43. St. Vincent's Hospital of New York, *Annual Report 1881, Annual Report 1889, Annual Report 1896, Annual Report 1900, Annual Report 1906;* Rosenberg, *Care of Strangers,* 147–49; Morris Vogel, *The Invention of the Modern Hospital: Boston, 1870–1930* (Chicago: University of Chicago Press, 1980), 60–61.

44. Rev. J. F. Richmond, *New York and Its Institutions 1609–1872* (New York: E. B. Treat, 1872), xiv–xv; New York City Department of Finance, *Private Charitable Institutions Receiving Public Money in New York City* (New York: Martin Brown Press, 1904), 15–32.

45. Walsh, *Sisters of Charity,* 3:99, 211–20, 225–27; Barbara Bates, *Bargaining for Life,* 173–96, discusses the development of private sanatoriums like these Catholic ones.

46. Rev. E. J. Crawford, *The Daughters of Dominic on Long Island,* 2 vols. (New York: Benziger Brothers, 1938, 1953), 1:188; "Activities of a Century in the United States," ts. (1958), n.p., Record Group 6, Series 6, FSP.

47. St. Vincent's Hospital of New York, *Annual Report 1881, Annual Report 1888, Annual Report 1894, Annual Report 1906.*

48. The economic and social sources of this change are examined in Rosenberg, *Care of Strangers;* David Rosner, *A Once Charitable Enterprise: Hospitals and Health Care in Brooklyn and New York, 1885–1915* (Cambridge: Cambridge University Press, 1982); and Vogel, *Invention of the Modern Hospital.*

49. *Metropolitan Record,* 16 April 1859.

50. *Metropolitan Record,* 26 April 1862.

## CHAPTER FOUR. "Building in New York Is Very Expensive"

1. St. Vincent's Hospital, *Thirty-Ninth Annual Report for the Year 1888,* 22v-2a, ASCMSV.

2. Sister Aloysia, Daughters of Charity Emmitsburg to Bernadette McCauley, 1 September 1989; Sisters of St. Joseph Brentwood, Annals, 1904, SSJB; Sister Marie de Lourdes Walsh, *The Sisters of Charity of New York,* 3 vols. (New York: Fordham University Press, 1960) 1:148.

3. James Gollin, *Worldly Goods: The Wealth and Power of the American Catholic Church, the Vatican, and the Men Who Control the Money* (New York: Random House, 1971), 99; Edward R. Kantowicz, *Corporation Sole: Cardinal Mundelein and Chicago Catholicism* (Notre Dame: University of Notre Dame Press, 1983), 34–35.

4. Kantowicz, *Corporation Sole,* 157; Gollin, *Worldly Goods,* 147–48.

5. In a successful effort to wrest control of New York's parishes from laymen, Archbishop Hughes deliberately restructured New York's parish corporations in the 1840s. He appointed himself and his successors the president of each; the diocesan vicar general was designated treasurer and the parish pastor, vice-president and secretary. While laymen remained on the parish boards, Hughes made sure they could always be outvoted by these permanent clerical members; Gollin, *Worldly Goods,* 99. Jay P. Dolan discusses the political and social implications of Hughes's reordering of New York parishes in *The Immigrant Church: New York's Irish and German Catholics, 1815–1865* (Notre Dame: University of Notre Dame Press, 1983), 64–65.

6. *Articles of Incorporation St. Joseph's of Far Rockaway,* 11 October 1905, St. Joseph's Hospital file, SSJB; Sisters of Charity of New York, Board Meetings 1847–1891, St. Mary's General Hospital, *Annual Report 1886,* St. Mary's folder; St. Vincent's Hospital, *Fourteenth Annual Report for the Year 1863, Thirty-Second Annual Report for the Year 1881, Thirty-Ninth Annual Report for the Year 1888, Forty-Second Annual Report for the Year 1891, Forty-Third Annual Report for the Year 1892, Forty-Fourth Annual Report for the Year 1893, Forty-Fifth Annual Report for the Year 1894, Forty-Sixth Annual Report for the Year 1895,· Forty-Seventh Annual Report for the Year 1896, Fifty-First Annual Report for the Year 1900, Fifty-Seventh Annual Report for the Year 1906:* 22v-2a, SCMSV; New York State Board of Charities, *Annual Report 1897,* 602–3; Walsh, *Sisters of Charity,* 3:212.

St. Elizabeth's had been founded in the 1870s as an old age home by third order Franciscans who ultimately opened a hospital. In the early 1890s, the Franciscan Sisters of Allegheny took over. New York State Board of Charities, *Annual Report 1897,* 602–3; Rev. Msgr. Florence D. Cohalan, *A Popular History of the Archdiocese of New York* (Yonkers: U.S. Catholic Historical Society, 1983), 308; John Gorrell, M.D., *St. Elizabeth's Hospital* (n.p.: Catholic Archdiocese of New York, 1951), 3. A third order community is a noncloistered group of men or women who live communally and are affiliated with a religious order. Evangeline Thomas, ed., *Women Religious History Sources: A Guide to Repositories in the United States* (New York: R. R. Bowker, 1983), xxvii.

7. *Hospital Progress* 2 (April 1921): 120.

8. Father Lewis to Mother Jerome, 19 July 1867, file 513M-17, SCMSV.

9. Sister Josepha Witzelhofer, to Ludwig Mission-Verein, 20 September 1858, copy with translation, SSDCHC.

10. Elinor Tong Dehey, *Religious Orders of Women in the United States* (Hammond, IN: W. B. Conkey, 1930), 224–25; Sister Mary Ignatius Meany, *By Railway or Rainbow: A History of the Sisters of Saint Joseph Brentwood* (Brentwood, NY: Pine Press, 1964), 58–62; Walsh, *Sisters of Charity* 1:114, 142–45, 2:269–71.

11. Sister M. Jerome to Sisters of Charity, Halifax, Letterbook, 24 May 1853, SCMSV.

12. St. Vincent's Hospital, *Annual Report 1863*; U.S. Department of Commerce, *Benevolent Institutions 1910* (Washington, DC: U.S. Government Printing Office, 1913).

13. *Brooklyn Tablet,* 17 October 1942, clipping, St. John's Hospital file, SSJB; Walsh, *Sisters of Charity,* 3:133.

14. St. Vincent's Hospital, "Journal of the Fair," 1882, file 522v-2c, SCMSV; Walsh, *Sisters of Charity,* 3:133–52.

15. St. Francis' Hospital, *Annual Report 1905,* AANY.

16. St. Vincent's Hospital, *Annual Report 1863*.

17. St. John's Hospital, *Annual Report 1891–2,* St. John's Hospital file; Sisters of St. Joseph's Brentwood, Annals, 1904–5, SSJB.

18. Both St. Luke's and Presbyterian hospitals, which also opened in this period, were founded with large individual donations. Albert Lamb, *The Presbyterian Hospital and the Columbia Presbyterian Medical Center* (New York: Columbia University Press, 1955), 8–18; James G. Wilson, *The History of the City of New York,* vol. 4 (New York: n.p., 1892–93), 438–40.

19. *Freeman's Journal,* 6 December 1856; *Metropolitan Record,* 12 October 1861; "Journal of the Fair"; St. Mary's General Hospital, *Annual Report 1886* and *St. Mary's Hospital Record and Guide to the Fair,* 2 December 1878, St. Mary's Hospital file, SCMSV. "Charities of NYC," *Catholic World,* September 1886, 810; Colleen McDannell, "Going to the Ladies' Fair, Irish Catholics in New York City, 1870–1900," in *The New York Irish,* ed. Ronald H. Bayor and Timothy J. Meagher (Baltimore: Johns Hopkins University Press, 1996), 237–43; Walsh, *Sisters of Charity,* 3:140.

20. *Freeman's Journal,* 6 December 1856.

21. Ibid.; St. Vincent's Hospital, *Annual Report 1863*; Walsh, *Sisters of Charity,* 3:139–40.

22. The reference to non-Catholic participation in the event was a not-so-subtle swipe explaining how Know-Nothing anti-Catholicism backfired to the advantage of the maligned. The article concluded by noting that the other Protestants had "been won towards the Catholic body by the atrocities of the

late Know-Nothing insurrection and its failures"; *Freeman's Journal,* 6 December 1856; Mother Seraphime Staimer, "Founding of the Convent of the Dominican Sisters (1853) at the Holy Trinity Church in Williamsburg with Its Dependent Convents," vol. 1 [ca. 1865], translated from German, 14, SSDCHC.

23. St. Vincent's Hospital, *Annual Report 1894*; *New York Times,* 20 December 1894; Walsh, *Sisters of Charity,* 3:329; Charles R. Morris, *American Catholic* (New York: Vintage Books, 1997), 10. Kelly's generosity was not limited to St. Vincent's Hospital. His will directed that $10,000 of his estate be directed toward "Hebrew Charities" in New York. With the approval of Archbishop Corrigan on this and other bequests, Beth Israel Hospital received part of the total amount Kelly designated. Edward Kelly to Archbishop Corrigan, 24 July 1895, AANY; Beth Israel Hospital, *Annual Report 1896,* NYAM.

24. Sister Thomas Francis Cushing, oral history, 21 March 1984, SSJB; Rev. E. J. Crawford, *Daughters of Dominic on Long Island,* 2 vols. (New York: Benziger Brothers, 1938, 1952), 1:253.

25. Morris J. Vogel, *The Invention of the Modern Hospital: Boston, 1870–1930* (Chicago: University of Chicago Press, 1980), 17.

26. Saint Mary's Hospital, *Annual Report 1917,* Saint Mary's Hospital file, SCMSV; Sr. M. Catherine Herbert to Sister M. Jeanette, 1940, Mary Immaculate Hospital file, SSDCHC.

27. New York State Charities Board, *Commissioner of Public Charities Annual Report 1869,* 216; "Claims of Fathers and the Congregation of the Church of the Most Holy Redeemer, New York, to the St. Francis' Hospital 5th Street New York," 20 July 1868, St. Francis' Hospital file, AANY.

28. St. Vincent's Hospital, *Annual Report 1863.*

29. St. Vincent's Hospital, *Annual Report 1888.*

30. Henry J. Camman and Hugh N. Camp, *The Charities of New York, Brooklyn and Staten Island* (New York: Hurd and Houghton, 1868), 49; St. Vincent's Hospital, *Annual Report 1895.*

31. St. John's Hospital Long Island City, *Annual Report 1891,* SSJB.

32. St. Vincent's Hospital, *Annual Report 1863, Annual Report 1881, Annual Report 1888, Annual Reports 1891–1896, Annual Report 1899, Annual Report 1906.*

33. St. Vincent's Hospital, *Annual Report 1863.*

34. St. Mary's Hospital, *Annual Report 1886.*

35. St. Vincent's Hospital, *Annual Report 1892*; St. Mary's Hospital, *Annual Report 1886.*

36. St. Vincent's Hospital, *Annual Report 1893.* For a lively discussion of some of the complications surrounding private bequests to religious organization, see Gollin, *Worldly Goods,* chap. 6.

37. May and the Dominicans reached a compromise whereby one-quarter of the monies collected by sisters at the convent affiliated with Saint Catherine's would remain the exclusive use of the Sisters for whatever purposes they wished. Staimer, "Founding of the Convent," 1:14; Sisters of Saint Joseph

Brentwood, Annals, July 1905, SSJB; Crawford, *Sisters of Dominic* 1:252, 257–60.

38. St. Mary's Hospital, *Annual Report 1886.*

39. St. Vincent's Hospital, *Annual Report 1893, 1881.*

40. St. Vincent's Hospital, *Annual Report 1895.*

41. *Freeman's Journal,* 14 April 1849; New York State Charities Board, *Annual Report 1903,* vol. 3, *Charity Legislation in New York 1609–1900.*

42. As noted earlier, the Sisters of Charity ran a military hospital in New York City. See Walsh, *Sisters of Charity,* 3:166–75.

43. John Webb Pratt, *Religion, Politics and Diversity: The Church State Theme in New York History* (Ithaca: Cornell University Press, 1967), 205–17. See also Paul Weber and Dennis A. Gilbert, *Private Churches and Public Money: Church-Government Fiscal Relations* (Westport, CT: Greenwood, 1981).

44. New York State Charities Board, *Annual Reports 1868–1870*; Pratt, *Religion, Politics and Diversity,* 205–17.

45. St. Vincent's Hospital, *Annual Report 1863*; Alexander B. Callow Jr., *The Tweed Ring* (New York: Oxford University Press, 1966), 82–83.

46. Pratt, *Religion, Politics and Diversity,* 211–12. The history of the construction of the wall of separation between church and state has been recently reexamined and explained in the context of nineteenth-century anti-Catholicism. See Philip Hamburger, *Separation of Church and State* (Cambridge, MA: Harvard University Press, 2002); John T. McGreevey, *Catholicism and American Freedom* (New York: W. W. Norton, 2003).

47. *Freeman's Journal,* 23 March 1850.

48. *Metropolitan Record,* 20 July 1867.

49. New York State, *Proceedings and Debates of the Constitutional Convention of the State of New York, 1867–1868* (Albany, 1868), as cited in Pratt, *Religion, Politics and Diversity,* 213.

50. Ibid., 210, 213–16, 219–20.

51. New York State Charities Board, *Annual Reports 1868–1870.*

52. Pratt, *Religion, Politics and Diversity,* 210–11.

53. "The Attack on Catholic Charities in New York," *Catholic World* 59 (October 1894): 702–9; Pratt, *Religion, Politics and Diversity,* 260.

54. "Private Charities and Public Money," *Catholic World* 29 (May 1879): 255–83; "Attack on Catholic Charities in New York," 702–9.

55. New York State Charities Board, *Annual Report 1879, Annual Report 1881.*

56. New York State Board of Charities, *Minutes,* 12 April 1905; Pratt, *Religion, Politics and Diversity,* 266.

57. As explained in n. 6, St. Elizabeth's Hospital was founded by a religious group of women who lived together and referred to themselves as sisters. In a very vivid example of ward boss politics and in a more direct manner than would be the case at other hospitals, the superior of St. Elizabeth's, Sister Mary Francis, wrote to Mayor Wickham in 1873 to "ask your aid (as a member of the board of apportionment) that we may continue in our good work and give

medical care and lodging to overwhelmingly poor people in the wards of the hospital." Significantly, her note was sent along to the mayor along with an addendum from John Kelly, New York's first Irish-Catholic city boss, who suggested that the mayor "look into this matter and do the best you can for them." Sister Mary Francis to Mayor Wickham, 19 December 1873; John Kelly to Hon. H. Wickham, December 1873, file 81-WW-2, NYCMA.

58. David Rosner, *A Once Charitable Enterprise: Hospitals and Health Care in Brooklyn and New York, 1885–1915* (Cambridge: Cambridge University Press, 1982), 122–33.

59. Ibid., 134–45.

60. Ibid., 161–63, 188–91.

61. "St. Vincent's Hospital," card, 1 May 1851, file 522-2b, SCMSV; *Freeman's Journal,* 22 June 1850.

62. St. Vincent's Hospital, *Annual Report 1863, Annual Report 1881, Annual Report 1888, Annual Report 1891–6, Annual Report 1900.*

63. New York State Charities Board, *Annual Report 1897,* 2:603.

64. Rosner, *A Once Charitable Institution,* 138.

65. Rosenberg, *Care of Strangers,* 240. See also Gail Farr Casterline, "Catholic Hospitals in Philadelphia," *Pennsylvania Magazine of History and Biography* (July 1984): 289–314.

66. St. Vincent's Hospital, *Annual Report 1900, Annual Report 1906.*

67. St. Vincent's Hospital, *Annual Report 1906.*

68. The Ladies Auxiliary of St. Vincent's Hospital, invitation [ca. 1900], file 522v-2b, SCMSV; St. Joseph's Hospital, *Annual Report 1906,* euchre ticket book, 3 February 1911, St. Joseph's Hospital file, SSJB; *New York Herald,* Brooklyn Supplement, 22 May 1904, clipping, Mary Immaculate Hospital file; "St. Catherine's Ladies Aid Society 1899–1900," photograph, St. Catherine's Hospital file, SSDCHC. On the subject of economic and social mobility among nineteenth-century Catholic immigrant groups, see Jay P. Dolan, *The American Catholic Experience: A History from Colonial Times to the Present* (Garden City, NY: Doubleday, 1985); Thomas Kessner, *The Golden Door: Italian and Jewish Mobility in New York City, 1880–1915* (New York: Oxford University Press, 1977); and Stephen Thernstrom, *Poverty and Progress: Social Mobility in a Nineteenth Century City* (Cambridge, MA: Harvard University Press, 1964).

69. St. Vincent's Hospital, *Annual Report 1906*; *New York Herald,* Brooklyn Supplement, 22 May 1904, Mary Immaculate Hospital File, SSDCHC.

70. Sisters of St. Joseph of Brentwood, *Annals,* 1906.

71. *"Lawn Fete and Bazaar" for the Benefit of St. Joseph's Far Rockaway on the Hospital Grounds,* 7–10 August 1907, St. Joseph's Hospital file, SSJB.

72. John Richmond, *New York and Its Institutions* (New York: E. B. Treat, 1871), 377; *Freeman's Journal,* 28 February 1857, 9 May 1857.

73. *Freeman's Journal,* 28 February 1857.

74. "Who Shall Take Care of Our Sick?" 54.

75. Susan Reverby, *Ordered to Care: The Dilemma of American Nursing* (Cambridge: Cambridge University Press, 1987), 60–65.

76. Bishop McDonnell to Sister Mary David, 31 January 1901, St. John's Hospital file, SSJB.

## CHAPTER FIVE. "Trust in God but Put Your Shoulder to the Wheel"

1. Holy Family Hospital, *Annual Report 1920*, Holy Family Hospital file; St. Mary's Hospital, *Annual Report 1913*, *Annual Report 1914*, *Annual Report 1917*, *Annual Report 1919*, file 513M, SCMSV.

2. On the Progressive reform movement and its impact on health and hospital care, see Kenneth Ludmerer, *Learning to Heal: The Development of American Medical Education* (New York: Basic Books, 1985); David Rosner, *A Once Charitable Institution: Hospitals and Health Care in Brooklyn and New York, 1885–1915* (Cambridge: Cambridge University Press, 1982); Paul Starr, *The Social Transformation of American Medicine* (New York: Basic Books, 1982); and Rosemary Stevens, *In Sickness and in Wealth: American Hospitals in the Twentieth Century* (New York: Basic Books, 1998). On the response of particular institutions, see Joan E. Lynaugh, "From Respectability Domesticity to Medical Efficiency: The Changing Kansas City Hospital, 1875–1920," and Edward C. Atwater, "Women, Surgeons and a Worthy Enterprise: The General Hospital Comes to Upper New York State," in *The American General Hospital: Communities and Social Context*, ed. Diane E. Long and Janet Golden (Ithaca: Cornell University Press, 1998), 21–42, 40–66.

3. Mother M. Concordia, "The General Superior and the Hospital," *Hospital Progress* 4 (August 1923): 295; E. H. Lewinski-Corwin, *The Hospital Situation in Greater New York* (New York: G. P. Putnam's Sons, 1924), 35–36.

4. Lewinski-Corwin, *Hospital Situation in Greater New York*, 213; *Hospital Progress* 11 (March 1930): 160–61.

5. In the early twentieth century the Catholic hierarchy encouraged the growth of Catholic women's colleges. As Mary Oates has explained, "Their growing interest in women's colleges arose from concern over the numbers [of Catholic women] flocking to normal schools and state colleges within their dioceses. To discourage this dangerous trend, they urged, and occasionally ordered, religious communities of women to open local colleges"; Mary J. Oates, C.S.J., "The Development of Catholic Colleges for Women, 1895–1960," *U.S. Catholic Historian* 7 (Fall 1988): 416.

6. "Conference of Supervisors of Nurses," *Hospital Progress* 1 (June 1910): 259.

7. Catholic Charities of the Archdiocese of New York, *Report of the Catholic Charities of the Archdiocese of New York, May 1, 1920–March 1, 1922* (hereafter all as Catholic Charities, *Report*), 71.

8. Susan Reverby, *Ordered to Care: The Dilemma of American Nursing* (Cambridge: Cambridge University Press, 1987), 68–70, 85–86.

9. Sister Mary Loretto, "Arrangement of Hours for Student Nurses," *Hospital Progress* 11 (June 1930): 21–22.

10. Barbara Melosh, *The Physician's Hand: Work, Conflict, and Culture in American Nursing* (Philadelphia: Temple University Press, 1982), 37.

11. Lewinski-Cronin, *Hospital Situation in Greater New York,* 212–13. Nationally, the number of graduate nurses in hospitals only replaced nursing students in the 1940s; Reverby, *Ordered to Care,* 188.

12. St. Vincent's Hospital, *Fifty-Eighth Annual Report for the Year 1907,* file 522v-2a, SCMSV. On nurses' training programs and hospital schools, see Reverby, *Ordered to Care,* chaps. 3–4, and Melosh, *Physician's Hand,* chap. 2.

13. University of the State of New York, *Ninth Annual Report of the State Education Department for the School Year Ending July 31, 1912* (Albany: New York State Education Department, 1913), 230.

14. On the early-twentieth-century nursing reform movement, see Melosh, *Physician's Hand,* chaps. 1–2, and Reverby, *Ordered to Care,* chaps. 7–9.

15. Nancy Tomes, "The Silent Battle: Nurse Registration in New York State, 1903–1920," in *Nursing History: New Perspectives, New Possibilities,* ed. Ellen Lagemann (New York: Teachers College Press, 1983), 111–14.

16. Saint Vincent's Hospital School of Nursing, *75 Years of Adventure in Nursing* (1967), author's collection.

17. Carol K. Colburn and Martha Smith, *Spirited Lives: How Nuns Shaped Catholic Culture and American Life* (Chapel Hill: University of North Carolina Press, 1999), 203.

18. Sister Hildegard to Most Rev. Archbishop Henry Moeller, D.D., 6 December 1916, copy in St. Peter's Hospital file, Record Group 6, Series: St. Peter's Hospital, FSP. On a different set of circumstances surrounding this same issue in another congregation, see Susan Carol Peterson and Courtney Ann Vaughan-Roberson, *Women with Vision* (Urbana: University of Illinois Press, 1988), 11–15. In that case, involving Presentation sisters in South Dakota, the community's constitution allowed all aspects of nurses training, but the local hierarchy forbade one group to do maternity work.

19. Sisters of the Poor of St. Francis, Annals St. Peter's Hospital, vol. 3, 1900–1947, 44; "Rules of the Sisters of the Poor of St. Francis in rendering assistance at gynecological and obstetrical operations," n.d., St. Peter's Hospital file, Record Group 6, Series: St. Peter's Hospital, FSP.

20. *Hospital Progress* 11 (March 1930): 160–61. Reprint from *American Journal of Nursing* 34 (November 1934): n.p., author's collection; Sister Marie De Lourdes Walsh, *The Sisters of Charity of New York,* 3 vols. (New York: Fordham University Press, 1960), 3:148–49.

21. Walsh, *Sisters of Charity,* 3:148–49.

22. Sister Margaret Quinn, "Directors of St. John's Long Island City Hospital School of Nursing," ts., n.d., St. John's Hospital file, SSJB; Sister Magda Marie to Rev. John Hunt, 21 July 1971, St. Catherine's Hospital file, SSDCHC.

23. Robert I. Gannon, S.J., *Up to the Present: The Story of Fordham* (Garden

City, NY: Doubleday, 1967), 124, 158; Andrew M. Greeley, *From Backwater to Mainstream: A Profile of Catholic Higher Education* (New York: McGraw-Hill, 1969), 152–53; Edward J. Power, *A History of Catholic Education in the United States* (Milwaukee: Bruce Publishing Co., 1958), 247.

24. Gannon, *Up to the Present,* 137; Ludmerer, *Learning to Heal,* 96.

25. Gannon, *Up to the Present,* 137; Ludmerer, *Learning to Heal,* 229.

26. Abraham Flexner, *Medical Education in the United States and Canada: A Report to the Carnegie Foundation for the Advancement of Teaching* (1910; reprint, Bethesda, MD: Science and Health Publications, n.d.), 100–103; Ludmerer, *Learning to Heal,* 163, 223.

27. As quoted in Stevens, *In Sickness and in Wealth,* 63. Illustrating the difference of opinion on this issue, Columbia University's affiliation with Presbyterian Hospital in New York came about only after the school was unsuccessful in earlier attempts to affiliate with Roosevelt Hospital. The administration at Roosevelt Hospital decided not to merge with Columbia because they were reluctant to direct the resources of their institution toward the physician's educational needs and away from what they viewed as their primary goal of charity and patient care; Ludmerer, *Learning to Heal,* 160–65.

28. Gannon, *Up to the Present,* 120–28, 160.

29. Ibid., 156–59; Yasha Levenson, *To Whom It May Concern* (privately printed, n.d.), 46, 59. Jewish students came to Fordham because of its location and because of their exclusion from other schools. On the quota system in American colleges and universities in the early twentieth century, see John Higham, *Send These to Me: Jews and Other Immigrants in Urban America* (New York: Athenaeum, 1975), and Marcia Graham Synott, *The Half-Opened Door: Discrimination and Admission at Harvard, Yale and Princeton, 1900–1970* (Westport, CT: Greenwood, 1979).

30. Charles E. Rosenberg, *The Care of Strangers: The Rise of America's Hospital System* (New York: Basic Books, 1987), 278–82; David Rosner, "Doing Well or Doing Good: The Ambivalent Focus of Hospital Administration," in *American General Hospital,* ed. Long and Golden, 157–70; Stevens, *In Sickness and in Wealth,* 68–75.

31. Haven Emerson, M.D., and Anna C. Phillips, *Survey of Catholic Hospitals: Report of St. Joseph's Hospital* (1923), 16, Saint Joseph's Hospital file, ASSJB.

32. Rosner, "Doing Well or Doing Good," 155–66. In 1924 Marquette opened a graduate program offering a master's degree for hospital executives, but the program closed in 1927; Robert J. Shanahan, S.J., *The History of the Catholic Hospital Association* (St. Louis: Catholic Hospital Association of the United States and Canada, 1965), 57–64.

33. Sister M. Domitilla, "Post Graduate Work for Sisters," *Hospital Progress* 2 (September 1921): 330–31.

34. Sister Marie Immaculate Conception, "The Education of the Hospital Sister," *Hospital Progress* 2 (September 1921): 330.

35. James C. Kennedy, M.D., "St. Catherine's Hospital, Brooklyn 1878–

1920"; Sister M. Idelphonse, "The Record Room"; S. H. De Coste, M.D., "Pathology and X-Ray Labs"; and Charles A. Gordon, "Department of Gynecology-Obstetrics at St. Catherine's," *Hospital Progress* 2 (April 1921): 117–32; "The Nurses' Home for St. John's Long Island City, New York," *Hospital Progress* 3 (March 1922): 90–91; *Hospital Progress* 3 (April 1922): 168; Edward F. Croker, "Fire Drills at Mary Immaculate Hospital, Jamaica, Long Island, N.Y.," *Hospital Progress* 7 (October 1926): 413. On the origins of the Catholic Hospital Association, see Coburn and Smith, *Spirited Lives,* 202; Shanahan, *Catholic Hospital Association,* 1–23; Stevens, *In Sickness and in Wealth,* 94–95; and Christopher J. Kauffman, *Ministry and Meaning: A Religious History of Catholic Health Care in the United States* (New York: Crossroad, 1995), 168–81.

36. Catholic Charities, *Report 1920–1922,* 76.

37. Ibid., 66, 69; Catholic Charities, *Report 1925,* 37; *Report 1930,* 41–42. See Rosner, *Once Charitable Institution,* 55–61, for a discussion of when and why other hospitals in New York began to charge patients.

38. Emerson and Phillips, *Report of St. Joseph's Hospital,* fig. 7.

39. Mother Joseph to Bayard L. Peck, 7 October 1920, 16 March 1921, 10 May 1921, St. Lawrence Hospital Copy Book 1920–1922; *Mary Immaculate Hospital Administrative Manual,* 1 July 1965, Mary Immaculate Hospital file, SSDCHC; Walsh, *Sisters of Charity,* 3:220. Sisters remained gentle but shrewd with regard to business matters. The Missionary sisters first interest payment due the Sisters of Charity did not arrive on schedule, prompting Mother Joseph to ask her attorney to nudge them. "We feel that a word from you would be more impressive than a note from us"; Mother Joseph to Bayard L. Peck, 16 May 1921, St. Lawrence Hospital Copy Book 1920–1922.

40. Thomas F. Daly, "How a Million Was Raised," *Hospital Progress* 6 (April 1925): 150–53.

41. Ibid., 153.

42. "St. Vincent's Hospital, New York City," *Hospital Progress* 7 (February 1922): 68.

43. Catholic Charities, *Report 1925,* 35–36.

44. Ibid.

45. Catholic Charities, *Report 1920–22,* 75; *Report 1926,* 46; *Report 1931,* 54.

46. Catholic Charities, *Report 1928,* 47.

47. Ibid.

48. John T. Hancock to Mother Joseph, 16 June 1919, St. Lawrence Hospital Letter Book, 2 April 1914, 16 July 1914, SCMSV; St. Joseph's Hospital, *Annual Report 1919*; Emerson and Phillips, *Report of St. Joseph's Hospital,* 16. For a discussion of how the increased reliance on a paying patient population affected a hospital's tax exempt status as a charitable institution, see Stevens, *In Sickness and in Wealth,* 41.

49. Catholic Charities, *Report 1928,* 46–47.

50. Emerson and Phillips, *Report of St. Joseph's Hospital,* fig. 8; George N.

Winkler, 15 December 1926, Annals of St. Joseph's Hospital, St. Joseph's Hospital file, SSJB.

51. Community Service Society, "Number of Protestants, Catholics, Hebrews, and Other Religious Affiliation of Patients Treated in the Fund Hospitals during 1925" (1925), n.p., Box 27, File 79-1, Community Service Society Collection, CUA.

52. Emerson and Phillips, *Report of St. Joseph's Hospital,* fig. 9, n.p.

53. Rev. C. A. Shyne, "What a Catholic Nurse Should Do for the Seriously Ill," *Hospital Progress* 8 (August 1924): 313.

54. Catholic Charities, *Annual Report 1931,* 55; Stevens, *In Sickness and in Wealth,* 142–43.

## EPILOGUE. "A Service So Dear"

1. Catholic Charities of the Archdiocese of New York, *Annual Report of the Catholic Charities of New York 1928,* 46–47.

2. Mother Mary Vincentia to the Sisters of Charity Mount St. Vincent, 30 September 1955, Holy Family Hospital file, SCMSV; Sister Marie de Lourdes Walsh, *The Sisters of Charity of New York,* 3 vols. (New York: Fordham University Press, 1960), 3:222.

3. Christopher J. Kauffman, *Ministry and Meaning: A Religious History of Catholic Health Care in the United States* (New York: Crossroads, 1995), 283; Rosemary Stevens, *In Sickness and in Wealth: American Hospitals in the Twentieth Century* (New York: Basic Books, 1989), 313, 333.

4. *A Guide to Health Services* (Our Lady of Mercy Medical Center, n.d. [ca. 1989]), n.p.; *Catholic Medical Center of Brooklyn and Queens* (New York: Catholic Medical Center of Brooklyn and Queens, n.d. [ca. 1989]), 2–3, author's collection; *The Official Catholic Directory* (New York: P. J. Kenedy and Sons, 1990); *New York Times,* 28 July 1990; Alliance for Catholic Hospitals and Human Services, *Our Outreach to the Community* (Catholic Health and Human Services, n.d. [ca. 1989]), n.p., SCMSV.

5. Felician A. Foy, O.F.M., *1990 Catholic Almanac* (Huntington, IN: Our Sunday Visitor, 1989); Jay P. Dolan, *The American Catholic Experience: A History from Colonial Times to the Present* (Garden City, NY: Doubleday, 1985), 438; George C. Stewart Jr., *Marvels of Charity: History of American Sisters and Nuns* (Huntington, IN: Our Sunday Visitor, 1994), 565.

6. Mother Mary Vincentia to Right Rev. Thomas E. Molloy, 1923, file 513M-18, SCMSV.

7. *Brooklyn Tablet,* 21 September 1940; *Nativity Mentor,* October 1940; *Brooklyn Citizen,* n.d., clippings, file 513M-18, SCMSV.

8. *A Fifty Year History of St. Mary's Hospital* (New York: St. Mary's Hospital, 1932), NYAM; Walsh, *Sisters of Charity,* 3:209–10.

9. Walsh, *Sisters of Charity,* 3:220–23.

10. Anthony J. J. Rourke, M.D., to Rev. Msgr. Joseph F. Brophy, 25 May

1954; Mother Mary Vincentia to Rev. Msgr. Joseph Brophy, 12 January 1955; Rev. Joseph Brophy to Mother Mary Vincentia, 25 January 1955; Rev. Msgr. Joseph F. Brophy to Mother Mary, 29 August 1955; Notes of Meeting, Sisters of Charity and Msgr. Brophy, 29 July 1955; Sisters of Charity Mount St. Vincent Minute Book, 30 July 1955, Holy Family Hospital file, SCMSV.

11. Mother M. Vincentia to Right Rev. Thomas E. Molloy, D.D., 6 June 1925; Rev. John M. Hilpert to Mother Josepha, 3 August 1926, Sisters of Charity Minute Book, July 1955.

12. Mother M. Vincentia McKenna to Right Rev. Thomas E. Molloy, D.D., 17 August 1922; Right Rev. Thomas E. Molloy, D.D., to Rev. John Hilpert, 13 May 1927; Rev. John Hilpert, Directives of the Right Rev. Bishop, May 1927, Sisters of Charity Minute Book, July 1955.

13. Sisters of Charity Minute Book, July 1955.

14. Sisters of Charity Minute Book, 30 July 1955.

15. Rev. Msgr. Joseph F. Brophy to Mother Mary, 29 August 1955, Sisters of Charity Minute Book, August 1955.

16. Rev. E. J. Crawford, *The Daughters of Dominic on Long Island,* 2 vols. (New York: Benziger Brothers, 1938, 1952), 2:135; William Jarvis, "Mother Seton's Sisters of Charity" (Ph.D. diss., Columbia University, 1984), 302.

17. Barbara Ferraro and Patricia Hussey with Jane O'Reilly, *No Turning Back* (New York: Poseidon Press, 1990), 22. Crawford, *Daughters of Dominic,* 2:242–43; Sister Mary Ignatius Meany, *By Railway or Rainbow: A History of the Sisters of Saint Joseph Brentwood* (Brentwood, NY: Pine Press, 1964), 246–48.

18. Dolan, *American Catholic Experience,* 425.

19. Ibid., 438.

20. Ferraro and Hussey, *No Turning Back,* 57–58.

21. Mary Ewens, O.P., "Women in the Covent," in *American Catholic Women: A Historical Exploration,* ed. Karen Kennelly, C.S.J. (New York: Macmillan, 1989), 33.

22. Mother M. Vincentia to Right Rev. Thomas E. Molloy, D.D., 6 June 1925; Anthony J. J. Rourke, M.D., to Rev. Msgr. Joseph Brophy, 25 May 1954; Rev. Msgr. Joseph F. Brophy to Mother Mary, 29 August 1955, Holy Family Hospital file, SCMSV.

23. Catholic medical schools founded in the nineteenth and early twentieth centuries that continued past the 1920s were at Georgetown, St. Louis, Creighton, Loyola, and Marquette Universities; Edward J. Power, *A History of Catholic Education in the United States* (Milwaukee: Bruce Publishing Company, 1958), 243–48.

24. *The Linacre Quarterly* 17 (May 1950): n.p.; Stevens, *In Sickness and Wealth,* 19, 112. Christopher Kauffman concludes that the National Federation of Catholic Physician's Guilds, founded in 1931 and headquartered in New York, was not a very significant force among physicians or in Catholic hospitals: "Only as strong as the numbers of local affiliates and based on volunteers among a profession subjected to long hours, the federation appears to have

been in a precarious position during its first ten years." Reorganized as the Association of Catholic Physicians in the 1940s, it never "came remotely close to representing a majority of Catholic physicians"; Kauffman, *Ministry and Meaning*, 237.

25. For a discussion of the relatively few numbers of Catholics among American physicians in the early twentieth century, see Kathleen Joyce, "Science and Saints: American Catholics and Health Care, 1880–1930" (Ph.D. diss., Princeton University, 1995), 188–96.

26. On the rise and then leveling off of the number of women physicians in the United States, see Regina Markell Moranz-Sanchez, *Sympathy and Science: Women Physicians in American Medicine* (New York: Oxford University Press, 1985), chaps. 4 and 9. The number of Catholic laywomen in medicine was limited too. In an analysis of Catholic laywomen in the national labor force between 1850 and 1950, Mary Oates concluded that the scarcity of Catholic women physicians reflected the fact that the economic status of the average Catholic family in the mid-nineteenth century was not conducive to the fulfillment of unusual career aspirations for daughters. By the 1920s, when greater numbers of Catholic women were acquiring means and motivation to pursue medical careers, the American Medical Association and the medical schools had imposed rigid and highly restrictive quotas on the admission of women. Mary J. Oates, C.S.J., "Catholic Laywomen in the Labor Force, 1850–1950," in Kennelly, *American Catholic Women*, 108.

27. One group of New York sisters, the Dominican Sisters of the Sick Poor, organized to nurse poor African Americans in their homes; Kauffman, *Ministry and Meaning*, 257.

# Bibliography

## Archives

The Archdiocese of New York, Saint Joseph's Seminary, Dunwoodie, Yonkers, NY 10704
Bellevue Hospital Archives, Bellevue Hospital, New York, NY 10016
Columbia University Archives, Butler Library, New York, NY 10017
Franciscan Sisters of the Poor Central Archives, 60 Compton Road, Cincinnati, OH 45215
Missionary Sisters of the Sacred Heart, Cabrini College, Eagle and King of Prussia Roads, Radnor, PA 19087
The New York Academy of Medicine, 2 East 103rd Street, New York, NY 10029
New York City Municipal Archives, 103 Chambers Street, New York, NY 10027
The New York Hospital, New York Hospital, New York, NY 10021
The New York Province of the Society of Jesus, Fordham University, Bronx NY, 10458
Redemptorist Provincial Residence, 7509 Shore Road, Brooklyn, NY 11209
The Sisters of Charity of New York, Boyle Hall, Mount Saint Vincent-on-Hudson, Riverdale, NY 10471
The Sisters of St. Dominic Congregation of the Holy Cross, 555 Albany Avenue, Amityville, NY 11701
The Sisters of St. Joseph Brentwood, New York, Saint Joseph Convent, 1725 Brentwood Road, Brentwood, NY 11717

## Books and Published Reports

Abell, Aaron I. *American Catholicism and Social Action*. Notre Dame, IN: University of Notre Dame Press, 1963.
Abel-Smith, Brian. *A History of the Nursing Profession in Great Britain*. New York: Springer, 1960.
———. *The Hospitals, 1880–1948: A Study in Social Administration in England and Wales*. Cambridge, MA: Harvard University Press, 1964.

Bates, Barbara. *Bargaining for Life: A Social History of Tuberculosis, 1876–1938.* Philadelphia: University of Pennsylvania Press, 1992.

Baxter, Annette, and Barbara Welter. *Inwood House.* New York: Inwood House, 1980.

Bayor, Ronald. H., and Timothy H. Meagher, eds. *The New York Irish.* Baltimore: Johns Hopkins University Press, 1996.

Beck, Sister Mary Bernice. *Handmaid of the Divine Physician.* Milwaukee: Bruce, 1952.

Bellah, Robert, and Frederick E. Greenspahn. *Uncivil Religion: Interreligious Hostility in America.* New York: Crossroad, 1987.

Betten, Neil. *Catholic Activism and the Industrial Worker.* Gainesville: University of Florida Press, 1976.

Billington, Ray Allen. *The Protestant Crusade, 1800–1860.* 1938. Reprint, Chicago: Quadrangle Books, 1964.

Bolster, Evelyn. *The Sisters of Mercy in the Crimean War.* Cork: Mercier Press, 1964.

Boylan, Marguerite T. *Social Welfare in the Catholic Church.* New York: Columbia University Press, 1941.

Brinton, John Hill. *Personal Memoirs.* New York: Neale, 1914.

Burgess, May Ayres. *Nurses, Patients, and Pocketbooks.* New York: Committee on the Grading of Nursing Schools, 1928.

Byrne, John F., C.SS.R. *The Redemptorist Centenaries.* Philadelphia: Dolphin Press, 1932.

Byrne, Sister Marie Le Gras. *A History of St. Vincent's Hospital School of Nursing.* Washington DC: Catholic University Press, 1941.

Callahan, Nelson, J. *The Diary of Richard L. Burtsell.* New York: Arno, 1978.

Cammann, Henry J., and Hugh N. Camp. *The Charities of New York, Brooklyn, and Staten Island.* New York: Hurd and Houghton, 1868.

Campbell, Helen Stuart. *Darkness and Light.* Hartford: Hartford Publishing Co., 1897.

Carlisle, Robert J., M.D., ed., *An Account of the Bellevue Hospital.* New York: Society of the Alumni of Bellevue Hospital, 1896.

Catholic Charities of the Archdiocese of New York. *Annual Reports.* New York: Catholic Charities, 1920–1940.

———. *Brief Summary of the Final Report of the Catholic Charities Diocesan Survey.* New York: Catholic Charities, 1920.

Clear, Caitriona, *Nuns in Nineteenth-Century Ireland.* Dublin: Gill and Macmillan, 1988.

Coburn, Carol K., and Martha Smith. *Spirited Lives: How Nuns Shaped Catholic Culture and American Life, 1836–1920.* Chapel Hill: University of North Carolina Press, 1999.

Cohalan, Rev. Msgr. Florence J. *A Popular History of the Archdiocese of New York.* Yonkers: New York Catholic Historical Society, 1983.

Colonial Society of Massachusetts. *Medicine in Colonial Massachusetts, 1620–*

*1820*. Boston: Colonial Society of Massachusetts; distributed by the University Press of Virginia, 1980.

Committee on the Grading of Nursing Schools. *Nursing Schools Today and Tomorrow.* New York: Committee on the Grading of Nursing Schools, 1934.

Cooper, Page. *The Bellevue Story.* New York: Thomas Y. Crowell, 1948.

Crawford, Rev. E. J. *The Daughters of Dominic on Long Island.* 2 vols. New York: Benziger Brothers, 1938, 1952.

Cray, Robert E., Jr. *Paupers and Poor Relief in New York City and Its Rural Environs, 1770- 1930*. Philadelphia: Temple University Press, 1988.

Cross, Robert. *The Church and the City.* Indianapolis: Bobbs-Merrill, 1967.

Cullen, Mary, ed. *Girls Don't Do Honors: Irish Women in Education in the Nineteenth and Twentieth Centuries.* Dublin: Argus Press, 1987.

Curran, Francis X., S.J. *The Return of the Jesuits.* Chicago: Lily University Press, 1966.

Curran, Robert Aimed, S.J. *Michael Augustine Corrigan and the Shaping of Conservative Catholicism in America.* New York: Arno, 1978.

Curtis, Sarah A. *Educating the Faithful: Religion, Schooling, and Society in Nineteenth-Century France.* Dekalb: Northern Illinois University Press, 2002.

Dearborn, Frederick M. *The Metropolitan Hospital.* New York: Privately printed, 1937.

Deferrari, Roy Joseph, ed. *Essays on Catholic Education in the United States.* 1942. Reprint, Freeport, NY: Books for Libraries, 1969.

Dehey, Eleanor Tong. *Religious Orders of Women in the United States.* Hammond, IN: W. B. Conkey, 1930.

Diner, Hasia. *Erin's Daughters in America: Irish Immigrant Women in the Nineteenth Century.* Baltimore: Johns Hopkins University Press, 1983.

Dolan, Jay P. *The American Catholic Experience: A History from Colonial Times to the Present.* Garden City, NY: Doubleday, 1985.

———. *The Immigrant Church: New York's Irish and German Immigrants, 1815–1865*. Notre Dame, IN: University of Notre Dame Press, 1983.

———. *In Search of American Catholicism.* New York: Oxford University Press, 2002.

Drinan, Robert F., S.J. *Religion, the Courts, and Public Policy.* New York: McGraw-Hill, 1963.

Dubos, Rene, and Jean Dubos. *The White Plague: Tuberculosis, Man, and Society.* 1953. Reprint, Brunswick: Rutgers University Press, 1987.

Ebaugh, Helen Fuchs. *Out of the Cloister.* Austin: University of Texas Press, 1977.

Ehrenreich, Barbara, and Deidre English. *Witches, Midwives, and Nurses.* Old Westbury, NY: Feminist Press, 1973.

Ellis, John Tracy, ed. *Documents of American Catholic History.* 3 vols. Wilmington, DE: Michael Glazier, 1987, 1956.

———. *Perspectives in American Catholicism.* Baltimore: Helicon, 1963.

Ellis, John Tracy, and Robert Trisco, eds. *A Guide to American Catholic History.* Santa Barbara, CA: ABC-Clio, 1982.

Emerson, Haven. *The Hospital Survey for New York.* 3 vols. New York: United Hospital Fund, 1937.

Ernst, Robert. *Immigrant Life in New York.* New York: Kings Crown Press, 1949.

Ewens, Mary, O.P. *The Role of the Nun in Nineteenth-Century America.* New York: Arno, 1978.

Ferraro, Barbara, and Patricia Hussey, with Jane O'Reilly. *No Turning Back.* New York: Poseidon Press, 1990.

Fichter, Joseph Henry. *Religion and Pain: The Spiritual Dimensions of Health Care.* New York: Crossroads, 1981.

Flexner, Abraham. *Medical Education in the United States and Canada: A Report to the Carnegie Foundation for the Advancement of Teaching.* 1910. Reprint, New York: Arno, 1972.

Francis, Samuel Ward. *Distinguished Living Surgeons.* New York: J. Bradburn, 1866.

Fraser, James W. *Between Church and State: Religion and Public Education in a Multicultural America.* New York: St. Martin's, 1999.

Fuller, Robert C. *Alternative Medicine and American Religious Life.* Oxford: Oxford University Press, 1989.

Gamble, Vanessa Worthington. *Making a Plan for Ourselves: The Black Hospital Movement, 1920–1948.* New York: Oxford University Press, 1995.

Gannon, Robert I., S.J. *Up to the Present: The Story of Fordham.* Garden City, NY: Doubleday, 1967.

Gavin, Donald. *The National Conference of Catholic Charities, 1910–1960.* Milwaukee: Bruce Press, 1962.

Gevitz, Norman, ed. *Other Healers: Unorthodox Medicine in America.* Baltimore: Johns Hopkins University Press, 1988.

Giles, Dorothy. *A Candle in Her Hand.* New York: G. P. Putnam's Sons, 1949.

Ginzberg, Lori D. *Women and the Work of Benevolence: Morality, Politics, and Class in the Nineteenth-Century United States.* New Haven: Yale University Press, 1990.

Goldie, Sue M. *"I Have Done My Duty": Florence Nightingale and the Crimean War, 1854–1856.* Iowa City: University of Iowa Press, 1987.

Gollaher, David L. *Voice for the Mad.* New York: Free Press, 1995.

Gollin, James. *Worldly Goods: The Wealth and Power of the American Catholic Church, the Vatican, and the Men Who Control the Money.* New York: Random House, 1971.

Gorrell, John, M.D. *St. Elizabeth's Hospital.* New York: Catholic Archdiocese of New York, 1951.

Granshaw, Lindsay, and Roy Porter, eds. *The Hospital in History.* London: Routledge, 1989.

Greeley, Andrew M. *The Catholic Experience: An Interpretation of the History of American Catholicism.* Garden City, NY: Doubleday, 1967.

———. *From Backwater to Mainstream: A Profile of Catholic Higher Education.* New York: McGraw-Hill, 1969.

———. *Irish Americans: The Rise to Money and Power.* New York: Harper and Row, 1981.

———. *An Ugly Little Secret: Anti-Catholicism in North America.* Kansas City, KS: Sheed Andrews and McMeel, 1977.

Griscom, John, M.D. *The Sanitary Condition of the Laboring Population of New York City.* New York: Harper and Brothers, 1845.

Grob, Gerald. *Mental Illness and American Society, 1875–1940.* Princeton, NJ: Princeton University Press, 1983.

Halsey, William. *The Survival of American Innocence.* Notre Dame, IN: University of Notre Dame Press, 1980.

Hamburger, Philip. *Separation of Church and State.* Cambridge, MA: Harvard University Press, 2002.

Harrar, James A., M.D. *The Story of the Lying-In Hospital of the City of New York.* New York: Society of the Lying-In Hospital, 1938.

Herron, Sister M. Eulalia. *The Sisters of Mercy in the United States.* New York: Macmillan, 1929.

Hickey, Edward J. *The Society for the Propagation of the Faith.* 1922. Reprint, New York: AMS, 1974.

Higham, John. *Send These to Me: Jews and Other Immigrants in Urban America.* New York: Athenaeum, 1975.

Hill, Michael. *The Religious Order.* London: Heinemann, 1973.

Hill, Sister M. Pauline. *In Love with Christ's Poor.* Cincinnati: St. Clare's Provincial House, 1959.

Hirsch, Joseph, and Beka Doherty. *The First Hundred Years of the Mount Sinai Hospital of New York.* New York: Random House, 1952.

———. *Saturday, Sunday, and Everyday: A History of the United Hospital Fund of New York.* New York: United Hospital Fund, 1959.

Hogan, Peter. *The Catholic University of America, 1896–1903.* Washington, DC: Catholic University of America Press, 1949.

Hudson, Winthrop S. *Religion in America.* 2nd ed. New York: Scribner, 1973.

Jacobs, Phillip P. *New Hospitals Needed in New York.* New York: New York State Charities Aid Association, 1908.

*The Jamaica Hospital, 1892–1942.* New York: Jamaica Hospital Medical Board, 1942.

James, Janet Wilson, ed. *Women in American Religion.* Philadelphia: University of Pennsylvania Press, 1980.

Jolly, Ellen Ryan. *Nuns of the Battlefield.* Providence: Providence Visitors Press, 1927.

Jones, Colin. *The Charitable Imperative.* London: Routledge, 1989.

Kantowitz, Edward R., ed. *American Catholicism, 1900–1965*. New York: Garland, 1988.

——. *Corporation Sole: Cardinal Mundelein and Chicago Catholicism*. Notre Dame, IN: University of Notre Dame Press, 1983.

Katz, Michael B. *Poverty and Policy in American History*. New York: Academic Press, 1983.

Kauffman, Christopher J. *Faith and Fraternalism: The History of the Knights of Columbus, 1882–1892*. New York: Harper and Row, 1982.

——. *Ministry and Meaning: A Religious History of Catholic Health Care in the United States*. New York: Crossroad Publishing, 1995.

Kennelly, Karen, C.S.J. *American Catholic Women: A Historical Exploration*. New York: Macmillan, 1989.

Kessner, Thomas. *The Golden Door: Italian and Jewish Mobility in New York City, 1880–1915*. New York: Oxford University Press, 1971.

Lagemann, Ellen, ed. *Nursing History New Perspectives, New Possibilities*. New York: Teachers College Press, 1983.

Lamb, Albert, M.D. *The Presbyterian Hospital and the Columbia Presbyterian Medical Center, 1868–1843*. New York: Columbia University Press, 1955.

Lannie, Vincent P. *Public Money and Parochial Education: Bishop Hughes, Governor Seward, and the New York School Controversy*. Cleveland: Press of Case Western University, 1968.

Leavitt, Judith Walzer. *Brought to Bed: Child Bearing in America, 1750–1950*. New York: Oxford University Press, 1986.

Leavitt, Judith Walzer, and Ronald Numbers, eds. *Sickness and Health in America: Readings in the History of Medicine and Public Health*. Madison: University of Wisconsin Press, 1978.

Lee, Eleanor. *Neighbors 1892–1967: A History of the Department of Nursing, Faculty of Medicine, Columbia University and Its Predecessor, the School of Nursing of the Presbyterian Hospital New York, 1892–1942*. New York: Columbia University–Presbyterian School of Nursing and School of Nursing Alumnae Association, 1967.

Levinson, Dorothy. *Montefiore: The Hospital as a Social Instrument 1884–1984*. New York: Farrar, Strauss, Giroux, 1984.

Levitan, Tina. *Island of Compassion: A History of Jewish Hospitals of New York City*. New York: Twayne, 1964.

Lewinski-Corwin, E. H. *The Hospital Situation in Greater New York*. New York: G. P. Putnam's Sons, 1924.

Long, Diane E., and Janet Golden, eds. *The American General Hospital: Communities and Social Context*. Ithaca, NY: Cornell University Press, 1989.

Ludmerer, Kenneth M. *Learning to Heal: The Development of American Medical Education*. New York: Basic Books, 1985.

Magray, Mary Peckham. *The Transforming Power of the Nuns: Women, Religion, and Cultural Change in Ireland, 1700–1900*. New York: Oxford University Press, 1998.

Maher, Mary Denis. *To Bind Up the Wounds: Catholic Sister Nurses in the Civil War.* New York: Greenwood Press, 1989.

Marshall, Helen E. *Dorothea Dix.* New York: Russel and Russel, 1937, 1967.

Massey, Mary Elizabeth. *Bonnet Brigades.* New York: Alfred A. Knopf, 1966.

McAvoy, Thomas. *A History of the Catholic Church in the United States.* Notre Dame, IN: University of Notre Dame Press, 1969.

McCarthy, Thomas P., C.S.V. *Guide to the Catholic Sisterhoods in the United States.* Washington DC: Catholic University of America Press, 1964.

McColgan, Rev. Daniel T. *A Century of Charity.* Milwaukee: Bruce, 1951.

McDannell, Colleen. *Material Christianity: Religion and Popular Culture in America.* New Haven: Yale University Press, 1995.

McGreevey, John T. *Catholicism and American Freedom.* New York: W. W. Norton, 2003.

McNamara, Jo Ann Kay. *Sisters in Arms: Catholic Nuns through Two Millennia.* Cambridge, MA: Harvard University Press, 1996.

Meany, Sister Mary Ignatius, C.S.J. *By Railway or Rainbow: A History of the Sisters of Saint Joseph Brentwood.* Brentwood, NY: Pine Press, 1964.

Melosh, Barbara. *The Physician's Hand: Work, Conflict, and Culture in American Nursing.* Philadelphia: Temple University Press, 1982.

Melville, Annabelle M. *Elizabeth Bayley Seton.* New York: Charles Scribner's Sons, 1951.

Mohl, Raymond A. *Poverty in New York, 1783–1825.* New York: Oxford University Press, 1971.

Morantz-Sanchez, Regina. *Conduct Unbecoming a Woman: Medicine on Trial in Turn-of-the-Century Brooklyn.* New York: Oxford University Press, 1999.

Morantz-Sanchez, Regina Markell. *Sympathy and Science: Women Physician's in American Medicine.* New York: Oxford University Press, 1985.

Morris, Charles R. *American Catholic.* New York: Vintage, 1997.

Mottus, Jane. *New York Nightingales: The Emergence of the Nursing Profession at Bellevue and New York Hospitals, 1850–1920.* Ann Arbor: UMI Research Press, 1980.

Nelson, Sioban. *Say Little, Do Much: Nursing, Nuns, and Hospitals in the Nineteenth Century.* Philadelphia: University of Pennsylvania Press, 2001.

Nightingale, Florence. *Notes on Nursing.* New York: Appleton, 1912.

Numbers, Ronald L., and Darrel W. Amundsen, eds. *Caring and Curing: Health and Medicine in the Western Religious Tradition.* New York: Macmillan, 1986.

Numbers, Ronald L., and Judith Walzer Leavitt. *Wisconsin Medicine: Historical Perspectives.* Madison: University of Wisconsin Press, 1981.

Oates, Mary J. *The Catholic Philanthropic Tradition in America.* Bloomington: Indiana University Press, 1995.

O'Brien, David. *American Catholics and Social Reform.* New York: Oxford University Press, 1968.

*The Official Catholic Directory.* Wilmette, IL: P. J. Kenedy and Sons, 1912–1936.

O'Grady, John. *Catholic Charities in the United States.* Washington DC: National Conference of Catholic Charities, 1931.

Opdycke, Sandra. *No One Was Turned Away: The Role of Public Hospitals in New York City since 1900.* New York: Oxford University Press, 1999.

Orsi, Robert. *The Madonna of 115th Street: Faith and Community in Italian Harlem, 1880–1950.* New Haven: Yale University Press, 1985.

——, ed. *Gods of the City: Religion and the Urban American Landscape.* Bloomington: Indiana University Press, 1999.

Osgood, Samuel. *New York in the Nineteenth Century.* New York: New-York Historical Society, 1866.

Patterson, James T. *The Dread Disease: Cancer and Modern American Culture.* Cambridge, MA: Harvard University Press, 1987.

Peterson, Susan Carol, and Courtney Ann Vaughn-Roberson. *Women with Vision: The Presentation Sisters of South Dakota, 1880–1985.* Urbana: University of Illinois Press, 1988.

Porter, Roy. *The Greatest Benefit Known to Mankind: A Medical History of Humanity from Antiquity to the Present.* New York: W. W. Norton, 1997.

Porterfield, Amanda. *Female Spirituality in America: From Sarah Edwards to Martha Graham.* Philadelphia: Temple University Press, 1980.

Power, Edward J. *A History of Catholic Higher Education in the United States.* Milwaukee: Bruce, 1958.

Pratt, John Webb. *Religion, Politics, and Diversity: The Church-State Theme in New York History.* Ithaca: Cornell University Press, 1967.

Rapley, Elizabeth. *The Devotes: Women and Church in Seventeenth-Century France.* Montreal: McGill-Queens University Press, 1990.

Reverby, Susan. *Ordered to Care: The Dilemma of American Nursing 1850–1845.* Cambridge: Cambridge University Press, 1987.

Reverby, Susan, and David Rosner, eds. *Health Care in America: Essays in Social History.* Philadelphia: Temple University Press, 1979.

Richmond, John F. *New York and Its Institutions, 1609–1871.* New York: E. B. Treat, 1871.

Risse, Guenter B. *Mending Bodies, Saving Souls: A History of Hospitals.* New York: Oxford University Press, 1999.

*The Roosevelt Hospital 1871–1957.* New York: Roosevelt Hospital Medical Board, 1957.

Rosenberg, Charles. *The Care of Strangers: The Rise of America's Hospital System.* New York: Basic Books, 1987.

——. *The Cholera Years: The United States in 1832, 1849, and 1866.* Chicago: University of Illinois Press, 1982.

——, ed. *Healing and History.* New York: Science History Publications, 1979.

Rosenberg, Charles, and Morris Vogel, eds. *The Therapeutic Revolution.* Philadelphia: University of Pennsylvania Press, 1979.

Rosenwaike, Ira. *Population History of New York City.* Syracuse: Syracuse University Press, 1972.

Rosenzweig, Roy, and Elizabeth Blackmar. *The Park and the People: A History of Central Park.* Ithaca: Cornell University Press, 1992.

Rosner, David. *A Once Charitable Enterprise: Hospitals and Health Care in Brooklyn and New York, 1885–1915.* Cambridge: Cambridge University Press, 1982.

Rothman, David J. *The Discovery of the Asylum.* Boston: Little, Brown, 1971.

Rothman, Sheila. *Living in the Shadow of Death: Tuberculosis and the Social Experience of Illness in American History.* New York: Basic Books, 1994.

Schoepflin, Rennie B. *Christian Science on Trial.* Baltimore: Johns Hopkins University Press, 2003.

Schultheiss, Karen. *Bodies and Souls: Politics and the Professionalization of Nursing in France, 1880–1922.* Cambridge, MA: Harvard University Press, 2001.

Shanahan, Robert. *The History of the Catholic Hospital Association, 1915–1965.* St. Louis: Catholic Hospital Association of the United States and Canada, 1965.

Sharp, John K. *History of the Diocese of Brooklyn, 1853–1953.* 2 vols. New York: Fordham University Press, 1954.

———. *Priests and Parishes of the Diocese of Brooklyn.* 2 vols. Brooklyn: Roman Catholic Diocese of Brooklyn, 1944, 1973.

Shaughnessy, Gerald S. *Has the Immigrant Kept the Faith.* New York: Macmillan, 1925.

Shaw, Richard. *Dagger John: The Unquiet Life and Times of Archbishop Hughes.* New York: Paulist Press, 1977.

Shea, John Gilmary. *The Catholic Church in New York City.* New York: L. C. Goulding, 1878.

———. *A History of the Catholic Church in the United States, 1843–1866.* 4 vols. New York: John G. Shea, 1886–1892.

Sister of St. Dominic (anon.). *History of the Catholic Church in the Diocese of Brooklyn.* New York: Benziger Brothers, 1938.

Smith, John Talbot. *The Catholic Church in New York.* 2 vols. New York: Hall and Locke, 1908.

Stansell, Christine. *City of Women: Sex and Class in New York City, 1789–1860.* New York: Alfred A. Knopf, 1986.

Starr, Paul. *The Social Transformation of American Medicine.* New York: Basic Books, 1982.

State Charities Aid Association of New York. *A Century of Nursing, with Hints toward the Organization of a Training School.* New York: G. P. Putnam's Sons, 1876.

Stepsis, Ursula, C.S.A., and Dolores Liptak, R.S.M., eds. *Pioneer Healers: The History of American Women in Health Care.* New York: Crossroad, 1989.

Stevens, Rosemary. *In Sickness and in Wealth: American Hospitals in the Twentieth Century.* New York: Basic Books, 1989.

Stewart, George C., Jr. *Marvels of Charity: History of American Sisters and Nuns.* Huntington, IN: Our Sunday Visitor, 1994.

Stuart, George Riley. *A History of St. Vincent's Hospital in New York City.* Privately printed, 1938.

Sullivan, Robert E., and James M. O'Toole, ed. *Catholic Boston: Studies in Religion and Community, 1870–1970.* Boston: Roman Catholic Archdiocese of Boston, 1985.

Sweet, William Warren. *Story of Religion in America.* New York: Harper and Brothers, 1930, 1939.

Synott, Marcia Graham. *The Half-Opened Door: Discrimination and Admission at Harvard, Yale, and Princeton, 1900–1970.* Westport, CT: Greenwood, 1979.

Taves, Ann. *The Household of Faith: Roman Catholic Devotions in Mid-Nineteenth-Century America.* Notre Dame, IN: University of Notre Dame Press, 1986.

Thernstrom, Stephen. *Poverty and Progress: Social Mobility in a Nineteenth-Century City.* Cambridge, MA: Harvard University Press, 1964.

Thomas, Evangeline, C.S.J., ed. *Women History Religious Sources: A Guide to Repositories in the United States.* New York: R. R. Bowker, 1983.

Tooley, Sarah A. *The History of Nursing in the British Empire.* London: S. H. Bousfield, 1906.

Tooley, Sarah A., and Bea Nergaard. *Ever Yours, Florence Nightingale: Selected Letters.* Cambridge: Cambridge University Press, 1990.

Vogel, Morris J. *The Invention of the Modern Hospital: Boston, 1870–1930.* Chicago: University of Chicago Press, 1980.

Walch, Timothy, ed. *Early American Catholicism, 1645–1820.* New York: Garland, 1988.

Walsh, Sister Marie de Lourdes. *The Sisters of Charity of New York 1809–1959.* 3 vols. New York: Fordham University Press, 1960.

———. *With a Great Heart: The Story of St. Vincent's Hospital and Medical Center of New York 1849–1964.* New York: St. Vincent's Hospital and Medical Center of New York, 1965.

Ward, David. *Poverty, Ethnicity, and the American City, 1840–1925: Changing Conceptions of the Slum and Ghetto.* Cambridge: Cambridge University Press, 1989.

Warner, John Harley. *The Therapeutic Perspective: Medical Practice, Knowledge, and Identity in America, 1820–1885.* Cambridge, MA: Harvard University Press, 1986.

Weber, Paul J., and Dennis A. Gilbert. *Private Churches and Public Money: Church Government Fiscal Relations.* Westport, CT: Greenwood, 1981.

Werner, M. R. *Tammany Hall.* New York: Greenwood, 1968.

White, Joseph, ed. *The American Catholic Religious Life.* New York: Garland, 1988.

White, William T. ed. *Medical Register of New York, New Jersey and Connecticut.* New York: G. P. Putnam's Sons, 1885.

Wilson, James G. *The Memorial History of the City of New York.* 4 vols. New York: New-York History Co., 1892–93.

## Newspapers and Periodicals

*Catholic World,* 1865–96
*The Freeman's Journal,* 1840–61
*Hospital Progress,* 1920–35
*The Linacre,* 1932–67
*Metropolitan Record,* 1864–68

## Articles

"The Attack on Catholic Charities in New York." *Catholic World* 59 (August 1884): 702–9.

Binsse, L. B. "Catholic Charities of New York." *Catholic World* 63 (September 1886): 809–20.

Browne, Henry J., ed. "The Archdiocese of New York a Century Ago: Memoir of Archbishop Hughes 1838–1858." *U.S. Catholic Historical Records and Studies* 39–40 (1952): 129–90.

Byrne, Patricia, C.S.J. "Sisters of Saint Joseph: The Americanization of a French Tradition." *U.S. Catholic Historian* 5 (Summer/Fall 1986): 241–72.

Casterline, Gail Farr. "St. Joseph's and St. Mary's: The Origins of Catholic Hospitals in Philadelphia." *Pennsylvania Magazine of History and Biography* 108 (1984): 289–314.

"Catholic Charities and the Microscope." *Catholic World* 60 (October 1894): 116–20.

Clarke, Richard H. "Catholic Life in New York City." *Catholic World* (May 1898): 192–218.

Cross, Robert. "Catholic Charities." *Atlantic Monthly* 210 (August 1962): 110–15.

Doyle, Ann, R.N. "Nursing by Religious Orders in the United States." *American Journal of Nursing* 29 (July–December 1929): 775–85, 959–69, 1085–95, 1197–1207, 1331–43, 1466–84.

Duncan, Richard R. "Masters' Theses and Doctoral Dissertations on Roman Catholic History on the United States: A Selected Bibliography." *U.S. Catholic Historian* 6 (Winter 1987): 51–114.

Dwight, Thomas, M.D. "Training Schools for Nurses of the Sisters of Charity." *Catholic World* 61 (May 1895): 187–92.

"The Early History of the Catholic Church on the Isle of New York." *Catholic World* 10 (December 1869): 413–20.

Elliot, Rev. Walter, C.S.P. "St. Vincent de Paul and the Sisters of Charity." *Catholic World* 70 (October 1899): 13–28.

Emery, S. L. "Charity Work of Catholic Women." *Catholic World* 68 (January 1899): 451–58.

Greeley, Andrew. "Catholics at the End of the Century." *Literary Review* 26 (Fall 1982): 5–11.

Hiestand, Wanda. "Nursing, the Family, and the 'New' Social History." *Advances in Nursing Science* 4 (April 1982): 1–11.

"Hospital Life in New York." *Harper's New Monthly Magazine* 57 (July 1878): 171–89.

Jamme, L. T. "Historical Sketch of the Society of St. Vincent de Paul in the United States, under the Jurisdiction of the Superior Council of New York, (written in 1884)." *U.S. Catholic Historical Society Records and Studies* 5 (1909): 195–209.

Kingdon, Frank. "Discrimination in Medical Colleges." *American Mercury* 61 (October 1945): 391–99.

Lewis, James R. "Mind-Forged Manacles: Anti-Catholic Convent Narratives in the Context of American Captivity Tale and Tradition." *Mid-America* 72 (October 1990): 150–67.

Locke, Clinton. "Founding a Hospital." *American Church Review* 33 (January 1881): 167–72.

Mannard, Joseph G. "Maternity of the Spirit: Nuns and Domesticity in Antebellum America." *U.S. Catholic Historian* 5 (Summer/Fall 1986): 305–24.

McCadden, Joseph, "Bishop Hughes versus the Public School Society." *Catholic Historical Review* 50 (1964): 188–207.

McDonough, J. V. "Catholic Schools and Charities under the New Constitution." *Catholic World* 62 (February 1896): 682–94.

Murphy, John T. "The Opportunities of Educated Catholic Women." *American Catholic Quarterly* (1898): 611–17.

"Notes on the Hospital of Warington, Florida." *Pensacola Historical Society Quarterly* 3 (July 1962): n.p.

Oates, Mary, C.S.J. "The Development of Catholic Colleges for Women, 1895–1960." *U.S. Catholic Historian* 7 (Fall 1988): 413–26.

O'Conner, James. "Anti-Catholic Prejudice." *American Catholic Quarterly Review* 1 (1876): 5–21.

O'Malley, Austin. "College Work for Catholic Girls." *Catholic World* 68 (November 1898): 161–67.

"Private Charities and Public Money." *Catholic World* 29 (May 1879): 255–83.

Quinn, Margaret, C.S.J. "Sylvia, Adele and Rosine Parmentier: 19th-Century Women of Brooklyn." *U.S. Catholic Historian* 5 (Fall 1986): 345–44.

Roosa, St. Johns, D.B., M.D., LL.D. "Proper Relation of Christians to Hospitals." *Christian Thought* 11 (1893–94): 382–90.

Rosenberg, Charles E. "And Heal the Sick." *Journal of Social History* 10 (June 1977): 428–47.

———. "Social Class and Medical Care in 19th-Century America: The Rise and Fall of the Dispensary." *Journal of the Social History of Medicine and the Allied Sciences* 29 (1974): 32–54.

"The Sanitary and Moral Condition of New York City." *Catholic World* 7 (July 1868): 553–66.

Schneider, Mary L., O.S.F. "American Sisters and the Roots of Change: The 1950's." *U.S. Catholic Historian* 7 (Winter 1989): 55–72.

Shea, John Gilmary. "The Rapid Increase of the Dangerous Classes in the United States." *American Catholic Quarterly* 4 (1879): 241–68.

Stevens, Rosemary. "A Poor Sort of Memory: Voluntary Hospitals and Government before the Depression." *Milbank Memorial Fund Quarterly* 60 (Fall 1982): 551–84.

Sullivan, Sister Mary Christine. "Some Non-Permanent Foundations of Religious Orders and Congregations." *U.S. Catholic Historical Society Records and Studies* 31 (1940): 117–18.

Thomas, J. Douglas. "A Century of American Catholic History." *U.S. Catholic Historian* 6 (Winter 1987): 22–51.

Thompson, Margaret Susan. "Discovering Foremothers: Sisters, Society, and the American Catholic Experience." *U.S. Catholic Historian* 5 (Summer/Fall 1986): 273–90.

Toner, J. M. "Statistics of Regular Medical Associations and Hospitals of the United States." *Transactions* 24 (1873): 285–333.

Walch, Timothy. "Catholic Social Institutions and Urban Development: The View for 19th-Century Chicago and Milwaukee." *Catholic Historical Review* 64 (January 1978): 16–32.

Weaver, MaryJo. "Feminist Perspectives and American Catholic History." *U.S. Catholic Historian* 5 (Summer/Fall 1986): 401–10.

Wilstach, Joseph Walter. "Frances Shervier and Her Poor Sisters." *Catholic World* 63 (May 1896): 261–64.

"Who Shall Take Care of Our Sick?" *Catholic World* 8 (October 1868): 42–55.

"Woman's Work in Religious Communities." *Catholic World* 58 (January 1894): 509–11.

## Government Documents and Reports

New York City Department of Finance. *Private Charitable Institutions Receiving Public Money in New York.* 1904.

New York City Inspector. *Annual Reports,* 1840–1858.

New York State Board of Charities. *Annual Reports.* 1868–1904.

———. *Directory of the Charitable, Eleemosynary, Correctional, and Reformatory Institutions of the State of New York.* 1892.

———. *Minutes of the Board Meeting.* 1867–98.

———. *Report of the State Board of Charities in Relation to Private Charity Institutions in New York City.* Assembly doc. 41, 1881.

New York State Metropolitan Health Board. *Annual Reports.* 1866–69.

University of the State of New York. *Ninth Annual Report of the Education Department for the School Year Ending July 31, 1912.* 1913.

———. *Tenth Annual Report of the Education Department for the School Year Ending July 31, 1913.* 1914.

U.S. Department of Commerce, Bureau of the Census. *Benevolent Institutions 1904.* Washington, DC: U.S. Government Printing Office, 1905.

———. *Benevolent Institutions 1910.* Washington, DC: U.S. Government Printing Office, 1913.

## Dissertations

Brown, Mary Elizabeth. "Italian Immigrants and the Catholic Church in the Archdiocese of New York, 1880–1950." Ph.D. diss., Columbia University, 1987.

Fitzgerald, Maureen. "Irish-Catholic Nuns and the Development of New York City's Social Welfare System 1840–1900." Ph.D. diss., University of Wisconsin–Madison, 1992.

Jarvis, William. "Mother Seton's Sisters of Charity." Ph.D. diss., Columbia University, 1984.

Joyce, Kathleen. "Science and the Saints: American Catholics and Health Care." Ph.D. diss., Princeton University, 1995.

Kelly, Mary. "Forty Shades of Green: Conflict over Community among New York's Irish, 1860–1920." Ph.D. diss., Syracuse University, 1997.

Liptak, Dolores Ann. "European Immigrants and the Catholic Church in Connecticut, 1870–1920." Ph.D. diss., University of Connecticut, 1979.

Maher, Mary Denis. "'To Do with Honor': The Roman Catholic Sister Nurse in the Civil War." Ph.D. diss., Case Western Reserve University, 1988.

Misner, Barbara. "A Comparative Study of the Members and Apostolates of the First Eight Permanent Communities of Women Religious within the Original Boundaries of the United States 1790–1850." Ph.D. diss., Catholic University, 1981.

Plattner, Elissa May. "How Beautiful upon the Mountains: The Sisters of the Divine Providence and Their Mission to Kentucky Appalachia." Ph.D. diss., University of Cincinnati, 1987.

Quiroga, Virginia. "Poor Mothers and Babies: A Social History of Childbirth and Child Care Institutions in Nineteenth Century New York City." Ph.D. diss., State University of New York at Stonybrook, 1984.

# Index

Page references in **boldface** type refer to illustrations.